New Directions
in Bioprocess Modeling
and Control:
Maximizing Process Analytical
Technology Benefits

New Directions in Bioprocess Modeling and Control:
Maximizing Process Analytical Technology Benefits

By

Michael A. Boudreau
Gregory K. McMillan

Notice

The information presented in this publication is for the general education of the reader. Because neither the author nor the publisher have any control over the use of the information by the reader, both the author and the publisher disclaim any and all liability of any kind arising out of such use. The reader is expected to exercise sound professional judgment in using any of the information presented in a particular application.

Additionally, neither the author nor the publisher have investigated or considered the affect of any patents on the ability of the reader to use any of the information in a particular application. The reader is responsible for reviewing any possible patents that may affect any particular use of the information presented.

Any references to commercial products in the work are cited as examples only. Neither the author nor the publisher endorse any referenced commercial product. Any trademarks or tradenames referenced belong to the respective owner of the mark or name. Neither the author nor the publisher make any representation regarding the availability of any referenced commercial product at any time. The manufacturer's instructions on use of any commercial product must be followed at all times, even if in conflict with the information in this publication.

Copyright © 2007
ISA—The Instrumentation, Systems, and Automation Society
All rights reserved.

Printed in the United States of America.
10 9 8 7 6 5 4 3 2

ISBN-13: 978-1-55617-905-1
ISBN-10: 1-55617-905-7

No part of this work may be reproduced, stored in a retrieval system, or transmitted in any form or by any means, electronic, mechanical, photocopying, recording or otherwise, without the prior written permission of the publisher.

ISA
67 Alexander Drive, P.O. Box 12277
Research Triangle Park, NC 27709
www.isa.org

Library of Congress Cataloging-in-Publication Data
Boudreau, Michael A.
   New directions in bioprocess modeling and control : maximizing process
analytical technology benefits / by Michael A. Boudreau, Gregory K.
McMillan.
       p. cm.
   Includes bibliographical references and index.
   ISBN 1-55617-905-7 (pbk. : alk. paper) 1. Biochemical engineering.
I. McMillan, Gregory K., 1946- II. Title.
   TP248.3.B68 2007
   660.6'3--dc22
                                                    2007026869

TABLE OF CONTENTS

Acknowledgments

This book would not have been possible without the enthusiastic support and commitment of resources by Mark Nixon and Grant Wilson of Emerson Process Management. In particular, the authors thank Grant Wilson for setting the path and establishing the perspective to achieve the full opportunities of a Process Analytical Technology initiative.

The authors express their appreciation to Joseph Alford of Eli Lilly and Company for his detailed and informative answers to our questions on current practice in the biopharmaceutical industry. This book tried to capture some of his extensive expertise in terms of the practical considerations in pursuing opportunities. The authors also express their appreciation to Harry Lam of Genentech, D. Grant Allen of the University of Toronto, and Nick Drakich of Invista (Canada) Company for their review of chapter 6 on first-principle modeling.

The authors thank the following people from Emerson Process Management for their technical contributions: Michalle Adkins (biopharmaceutical applications), Terry Blevins (basic and advanced control), Bruce Campney (demos), Ashish Mehta (neural networks), Dirk Thiele (model predictive control), Peter Wojsznis (adaptive control), and Willy Wojsznis (model predictive control). The authors thank Robert Heider from Washington University in Saint Louis and Martha Schlicher from Renewable Agricultural Energy, Inc., for contributions to the application of neural-network modeling in ethanol production. The authors thank Thomas Edgar and Yang Zhang from the University of Texas for their investigation and survey of principal component analysis for batch fault detection and analysis.

The kinetics and operating conditions in chapter 6 for the first-principle modeling of the production of ethanol from cellulose were developed by Nathan Mosier and Michael Ladisch in the Laboratory of Renewable Resources Engineering at Purdue University.

A portion of chapter 7 is extracted from a thesis presented by Michael N. May to the Henry Edwin Sever Graduate School of Washington University in Saint Louis in partial fulfillment of the requirements of the degree of Master of Science, 1 July, 2005.

Appendix E is an excerpt from an article by Michael J. Needham titled *"Enzyme Decreased by Controlling the pH with a Family of Bezier Curves"* published in the May 2006 issue of *Pharmaceutical* magazine.

Finally, the authors express their gratitude to Jim Cahill and Brenda Forsythe for the cover graphics.

About the Authors

Michael A. Boudreau, P.E., is a control systems engineer on the final leg of a multi-year, world-wide tour of distributed control system manufacturers; testing and developing advanced control systems on Emerson Process Management's DeltaV. During his career, Michael has worked on bioreactor control systems ranging in scale from 10 liter pharmaceutical pilot fermentors to 100,000 liter bulk chemical fermentors. He helped design the first EPO and G-CSF process control systems for Amgen, managed the control group at a Miles Labs (Bayer) citric acid plant and designed, started and maintained bacterial and cell culture fermenter control systems at Genentech.

Gregory K. McMillan, CAP, is a retired Senior Fellow from Solutia and Monsanto where he worked in engineering technology on process control improvement. Greg was also an affiliate professor for Washington University in Saint Louis. Presently, Greg is a contract consultant for Emerson Process Management in DeltaV R&D. Greg is an ISA Fellow and received the ISA "Kermit Fischer Environmental" award for pH control in 1991, the *Control* magazine "Engineer of the Year" award for the Process Industry in 1994, was inducted into the *Control* "Process Automation Hall of Fame" in 2001, and honored by *InTech* magazine in 2003 as one of the 50 all time most influential innovators in automation.

Preface

The Federal Drug Administration (FDA) website (www.fda.gov/cder/OPS/PAT.htm#Introduction) offers the following key points as an introduction to Process Analytical Technology.

Process Analytical Technology:

- A system for designing, analyzing, and controlling manufacturing through timely measurements (i.e., during processing) of critical quality and performance attributes of raw and in-process materials and processes with the goal of ensuring final product quality.

- It is important to note that the term *analytical* in PAT is viewed broadly to include chemical, physical, microbiological, mathematical, and risk analysis conducted in an integrated manner.

Process Analytical Technology Tools:

- There are many current and new tools available that enable scientific, risk-managed pharmaceutical development, manufacture, and quality assurance. These tools, when used within a system can provide effective and efficient means for acquiring information to facilitate process understanding, develop risk-mitigation strategies, achieve continuous improvement, and share information and knowledge. In the PAT framework, these tools can be categorized as:

    - Multivariate data acquisition and analysis tools

    - Modern process analyzers or process analytical chemistry tools

    - Process and endpoint monitoring and control tools

    - Continuous improvement and knowledge management tools

    - An appropriate combination of some, or all, of these tools may be applicable to a single-unit operation, or to an entire manufacturing process and its quality assurance.

This book is dedicated to synergistic knowledge discovery from the application of model predictive control, virtual plants, first-principle models, neural networks, and principal component analysis. This book seeks an integration of these tools in an environment for innovation that is the heart of PAT initiatives.

An important perspective in the implementation process is that process control transfers variability from its controlled variable to its manipulated variable. When you tune a loop for tight control, you see a straight line for the controlled variable but see fluctuations in the manipulated variable on a long-term trend. Process control does not make variability disappear. You cannot keep a controlled variable and manipulated variable both constant. In advanced control, you are transferring variability from a higher level process variable, such as product formation rate, to a lower level process variable such as substrate concentration. The level of control and transfer of variability has a profound impact on the choice of variables for neural network and multivariable statistical process control.

This book is a "quest for insight". Detailed insights are offered in each chapter. The following two general insights are valuable for orienting the reader for new directions in modeling and control of bioprocesses.

PAT is not just about analyzers. It is not process analyzer technology. It is process analytical technology.

Process control transfers variability. It doesn't make variability disappear.

Chapter 1:
Opportunities

Chapter 1

Opportunities

1-1. Introduction

The cell is the ultimate reactor. It is the essential and in some cases the only possible route to the production of complex compounds [1]. Bioreactors (fermenters) are the key unit operation in biopharmaceutical, brewing, biochemical, biofuel, and activated sludge processes. Each bioreactor relies on the performance of billions of these individual reactors or cells. Process control attempts to influence the sophisticated metabolic reactions inside the cell by controlling the environment immediately outside of the cell [2].

> *Process control attempts to influence the individual sophisticated internal reactions of billions of cells by controlling their extracellular environment.*

In order to control a process variable, we must measure it directly or infer it from other measurements. The key process variables of the cells' environment that are measured are compositions and conditions, such as temperature, pH, dissolved oxygen and carbon dioxide, and substrate. In process control, the compositions and conditions in the broth or vent gas are termed *process outputs* and the quantity, composition, and condition of feeds or charges into the bioreactor or of coolant into coils or jacket are termed *process inputs*. Seed cultures, nutrients, substrates, air, and oxygen are process inputs. In basic feedback control, process outputs that are measured or inferred are controlled by manipulating the process inputs. The goal of process control is to transfer variability from important process outputs to the process inputs designed to be manipulated. How well the process outputs are controlled is determined by loop dead time, measurement resolution, repeatability, noise, the tuning of the controllers, and the resolution of the control valves and variable speed drives [3].

Section 1-2 of this chapter discusses the major sources of variability, the definition of process inputs and outputs, the availability of measurements, the differences between experimental and first-principle models, the quality and quantity of data, and the different modes of batch operation. Section 1-3 discusses the different levels of control and the effect each has on the selection of model inputs and on the setup of basic feedback loops for different types of cell cultures. Section 1-4 introduces the important topics of online yield and capacity performance indicators. Section 1-5 outlines the use of model predictive control for optimization. Section 1-6 summarizes the reasons behind, the drivers of, and the tools used for the Process Analytical Technology (PAT) initiative.

Models offer benefits before they are put on line. It is the authors' experience that significant improvements result from the process knowledge and insight that are gained when building the experimental and first-principle models for process monitoring and control. The benefits often come in ways not directly attributable to the associated technologies, such as changes in batch set points, end points, and phases.

> *Significant improvements result from the process knowledge and insight that are gained when building the experimental and first-principle models for process monitoring and control.*

Doing modeling in the process development and early commercialization stages is advantageous because it increases process efficiency and provides ongoing opportunities for improving process control. When bench-top and pilot plant systems use the same industrial control systems and configuration expertise that are employed in manufacturing, applications of modeling and control can be developed as an integral part of the process definition and ported for industrial production via the control definition. The advanced technologies discussed in this book—model predictive control (chapter 4), the virtual plant (chapter 5), first-principle models (chapter 6), neural networks (chapter 7), and multivariate statistical process control (chapter 8)—are important tools for maximizing the benefits from process analyzers and tools. The synergistic discovery of knowledge is consistent with the intent behind the Process Analyzer and Process Control Tools sections of the FDA's "Guidance for Industry PAT – A Framework for Innovative Pharmaceutical Development, Manufacturing, and Quality Assurance" document, as the following excerpts show.

> *The synergistic knowledge discovered through the integration of analyzers, models, and controls is the essence of the process analytical technology (PAT) opportunity.*

In the Process Analyzer section of the FDA's guidance document we read:

> "For certain applications, sensor-based measurements can provide a useful process signature that may be related to the underlying process steps or transformations. Based on the level of process understanding these signatures may also be useful for the process monitoring, control, and end point determination when these patterns or signatures relate to product and process quality."

In the Process Control Tools section of the guidance document we read:

"Strategies should accommodate the attributes of input materials, the ability and reliability of process analyzers to measure critical attributes, and the achievement of process end points to ensure consistent quality of the output materials and the final product. Design and optimization of drug formulations and manufacturing processes within the PAT framework can include the following steps (the sequence of steps can vary):

- Identify and measure critical material and process attributes relating to product quality

- Design a process measurement system that allows real-time or near real-time (e.g., on-, in-, or at-line) monitoring of all critical attributes

- Design process controls that provide adjustments to ensure all critical attributes are controlled

- Develop mathematical relationships between product quality attributes and measurements of critical material and process attributes

Within the PAT framework, a process end point is not a fixed time; rather it is the achievement of the desired material attribute.... Quality decisions should be based on process understanding and the prediction and control of relevant process/product attributes."

The discovery and implementation of more optimal batch profiles, end points, and cycle times are encouraged by the PAT guidance document. The proper integration of technologies can actually speed up rather than slow down the time to market. Ideally, the optimization process is coincident with the commercialization process. Waiting until a pharmaceutical is out of patent protection increases the risk of not having the lowest cost position. The return on investment (ROI) gained by optimizing the production process for a new drug can be enough to justify the use of analyzers, models, and control. The goal of this book is to provide the basis for taking advantage of this opportunity.

Making the optimization process coincident with the commercialization process can lead to a low-cost position, an important factor for a competitive advantage when a drug goes off of patent protection.

Learning Objectives

A. Be able to track down the major sources of variability.

B. Recognize how instruments aggravate or mitigate variability.

C. Understand how to reduce variability through different levels of control.

D. Know the basic setup of loops for different types of bioreactors.

E. Appreciate the implications of missing online and infrequent lab measurements.

F. Become familiar with the requirements and capabilities of various models.

G. Understand how the level of control affects the selection of model inputs.

H. Discover optimization opportunities.

I. Be aware of the business drivers for the Process Analytical Technology (PAT) Initiative.

1-2. Analysis of Variability

In the pulp and paper industry, poor feedback control actually increases the variability of important process outputs. This phenomenon is rare for bioreactors because of the large volume, the slowness of the kinetics, and the lack of interactions. However, the actual performance the bioreactors achieve still depends on how well controllers are set up and tuned to deal with the sources of variability. Most feedback control implementation problems involving bioreactors can be traced to nonrepresentative measurements or mechanical design limitations in equipment, agitation, piping, and injection.

The bulk velocity in a bioreactor for animal cell cultures is extremely low and ranges from 0.01 to 0.1 ft/sec (scale of agitation ranges from 0.1 to 1.0, and power input per unit volume ranges from 10 to 100 watts/cubic meter). For bioreactors with fungi and bacterial cultures, the degree of agitation is comparable to chemical reactors with gas dispersion: bulk velocity is 0.8 to 1.0 (ft/sec) or larger (scale of agitation ranges from 8 to 10 and power-per-unit volume is 1000 watts/cubic meter or more) [4] [5] [6]. For ethanol, the scale of agitation ranges from 1 to 2 in the bioreactor and 2 to 4 in the mix tanks.

Low fluid velocities result in larger mixing time delays, less broth uniformity and gas dispersion, and lower mass transfer rates. The response time of a clean electrode and thermowell is several times larger in an animal cell culture. An even more important consideration is the increased propensity for coatings caused by low fluid velocity. A 1 millimeter coating on a pH measurement electrode can cause its response time to increase from 10 seconds to 7 minutes [7]. Chapter 2 on process dynamics discusses how these time delays and time lags add up to a total loop dead time that determines the loop's ultimate performance.

A coating on the reference electrode junction can cause a drift of 0.5 pH or more during the batch. Since the peak in the growth rate of the biomass with pH is relatively sharp, the pH drift can affect batch performance. Even if electrodes could be withdrawn, cleaned, and calibrated during a batch, the result would be disruptive because the time required for equilibration of the reference junction in the broth is significant [7]. Thus, even if the optimum pH does not change during the batch, there is an opportunity to counteract a drift in measurement by slowly trimming the pH set point. In continuous processes, it is quite common for operations to home in on a more optimum set point so as to compensate for sensor offset. However, in batch processes, the practice is to use fixed set points.

The low fluid velocities in animal cell bioreactors increase the total loop dead time and the need for optimizing the pH set point.

Load cells or mass inventories are frequently used as a standard way of adding some materials. Many believe that load cells and mass inventories are accurate standard technology, but the reality is that they are often the source of significant variability. To use load cells you must be sure that all piping and other connections to the vessel offer no vertical forces on the vessel (this includes operators who lean on the vessel when looking through the tank site glass). This is often difficult to achieve in reactors, since many steam sterilization and air/nutrient feed lines are connected to them. Also, changing ambient temperature conditions causes metal fittings and/or support structure to expand and contract and temperatures in the jacket cooling medium to change, which then changes the medium's density, hence weight, and so on. These changes can potentially change a load cell measurement unless the load cell is carefully engineered not to be affected or the change is adjusted for by correcting the model [8].

Poor control loop performance in bioreactors can often be traced back to nonrepresentative measurements or mechanical design limitations.

Totalized Coriolis mass flowmeters are used instead of load cells to achieve higher accuracy in charge measurements. These meters also provide an extremely precise density measurement for verifying the composition of charges. In the case of ethanol production, the density measurement on the feed to the fermenter provides an inference of the percentage of starch and sugar solids. Adding both a differential pressure measurement across the meter for laminar flow and a neural network produces a viscosity measurement that is an inference of starch-to-sugar conversion from the liquefaction process. Accurate mass flow measurements also facilitate an accurate mass flow ratio of feed to recycle flow at various points in the ethanol process.

A higher level of control, termed *optimization*, occurs when cell growth and product formation rates, which are indicative of intracellular reaction rates, are inferred from oxygen uptake rates (OUR) and carbon dioxide evolution rates (CER). Mass spectrometers can be used to provide online indications of OUR and CER by way of measuring the oxygen and carbon dioxide in the off-gas [1] [10]. Alternatively, the air or oxygen feed can be momentarily shut off and the rate of change in the dissolved oxygen used to infer the OUR. For ethanol production, the simple addition of a pressure measurement at the bottom of the fermenter enables us to infer CER from the broth's loss in weight. These rates can be potentially made more consistent and maximized at various points in the batch by adjusting the set point for key process outputs, such substrate concentration, dissolved gas concentration, and pH [8] [9]. In most cases, substrate concentration is not measured, and substrate feed (a process input) is set directly.

Most of the variability in the product or effluent can be tracked down to unknown variability in the process inputs. Measurable changes in seed cells are reduced by not transferring the seed tank until appropriate criteria are met. The process inputs that vary the most and with the greatest consequences are the unknown differences in the internal composition and metabolic pathways of cells. It is important to remember that a cell is the most sophisticated reactor known, and genetic engineering is still in its early stages. Fortunately, a virtual plant can achieve its objectives using relatively simple expressions for the kinetics and mass transfer rates, with the parameters fit to experimental data and combined with equations for the material, charge, and energy balances. We describe these procedures in chapter 6 on first-principle models.

The causes of variability in the seed cells are analogous to the causes of variability in people: the current and historical environment of the culture and genetics (heredity) [8]. In the case of environment, we have to consider the entire environment that the cell culture has been exposed to since the cell culture began as a few frozen cells in a liquid nitrogen storage container. These cells must be carefully thawed in a controlled fashion, and then go through several growth stages in different progressively larger equipment sets, typically lasting many days. During these stages the cells must be given all the appropriate nutrients and be controlled at the appropriate temperature, pH, and the like—and all of this performed in a sterile environment.

These "early" steps are typically not as well controlled (i.e., they require more manual operations) as the fermenter itself because more manual operations are entailed. So the opportunity for excursions, operator mistakes, that is, for variability, is significant [8].

There is a low, but finite, probability that a cell will mutate while dividing to become two cells. With each succeeding generation of cells, the probability increases that one or more of the cells has mutated. Note that by the time a culture is making product in a fermenter, perhaps thirty or more generations have occurred since the start of a few frozen cells. A mutated cell, theoretically, might produce the same or more product than a standard cell, but, in most cases, it produces less (or perhaps none). Companies specifically try to develop industrial processes so as to limit the number of cell generations because this minimizes concerns regarding cell mutations. The Food and Drug Administration (FDA) sometimes checks on this during its reviews. Progressive mutation is one of the reasons why bioprocesses are not typically run continuously (i.e., for very long periods of time) to maximize the probability that the culture will remain in a relatively pure state [8].

This is not to say that other bioreactor inputs cannot cause significant variability. The use of complex raw materials as nutrient sources, such as soybean meal or fish meal, is one of the most obvious significant sources of variability. This is particularly the case when a process shifts from one ocean fish harvest or soybean harvest to another. However, these kinds of changes cause sudden shifts in the productivity of all bioreactors making a product. The changes in nutrients align with any new lot of raw material being used rather than varying in a random lot-to-lot fashion, which might be associated with instrument noise or drift [8].

Process Inputs and Outputs
Figure 1-2a shows many of the process inputs and outputs that are possible for a bioreactor. Process inputs are typically quantities of reagents, substrates, nutrients, gases, enzymes, cells, or surfactants added to the vessel and cooling water circulated through the jacket. The flows of these process inputs are set by a final element, such as a pump or fan speed. Another type of process input is agitation set by motor speed. There are also unknown and often undesirable inputs such as impurities and contaminants. The process inputs determine the process outputs, such

as broth and jacket temperature, pressure, pH, dissolved gases, broth composition, and off-gas composition and flow [1] [10].

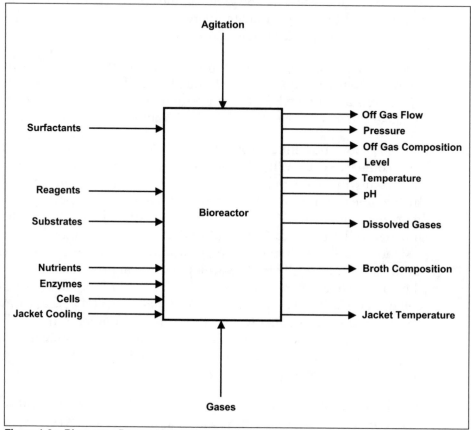

Figure 1-2a. Bioreactor Process Inputs and Outputs

Field and Laboratory Measurements

Even though the process outputs are indications of bioreactor performance, only the simpler measurements, such as temperature, pressure, pH, and differential pressure are typically installed on bioreactors. This is particularly true in the brewing and biofuel industries. The next most common measurement is dissolved oxygen for aerobic processes. Dissolved carbon dioxide electrodes have significantly improved since their appearance in the 1980s and are increasingly being used. In particular, their use is increasing for animal cell cultures when the pH is controlled by adding carbon dioxide, for the aeration and stripping of carbon dioxide in the broth by an air sparge, or for the sweeping of carbon dioxide from the head space by an inert gas [1]. Companies seeking higher levels of control add "at line" measurements of both components, such as glucose probes for substrate concentration and optical density instruments for biomass concentration.

Pharmaceutical manufacturers take advantage of sophisticated types of off-gas analyses on line. These analyses are generated by sending gas sample streams through field mass spectrometers that have a much better resolution, repeatability, reliability, and response time than other analyzers. Microprocessors and software have made possible diagnostics, computation capability, and user interfaces that greatly improve the mass spec's functionality and maintainability [1] [10].

Bioreactors do not have enough field measurements of process outputs.

The most important process output, broth composition, is measured in the laboratory. When this measurement is done in a laboratory, the time delay between the time when the sample was taken and when the analysis result is available to the control system is large and inconsistent. So too is the time interval between data points. The infrequency and variability in the timing of laboratory analysis makes it difficult to correlate process outputs with process inputs and to construct experimental models, such as partial least squares (PLS) and artificial neural network (ANN) models. This infrequency and timing variability also makes it nearly impossible to use the results for basic control directly. Inferential measurements (virtual sensors), which are discussed here and in chapter 5 on the virtual plant can provide rapid as well as more reliable measurements with less noise in broth composition, such as sugar and biomass concentration.

The laboratory measurement data of the most important process output, broth composition, is so sparse and delayed that conventional PLS and ANN modeling is difficult, making it unlikely that the lab data can be used directly for basic process control.

Experimental Models
Process models can be used to prevent abnormal situations, to detect bad batches, to experiment (exploring "what-if scenarios"), to prototype control systems, to check out configurations, to train personnel in operations, to predict process performance (batch end point, yield, and/or cycle time), and to optimize in real time.

Experimental models correlate a series of process outputs to process inputs. The three major types of experimental models used in industry (model predictive control [MPC], ANN, and PLS) are discussed in detail in chapters 4, 7, and 8, respectively. Some important fundamental conceptual differences distinguish these three models.

Artificial neural network (ANN) models predict a process output as a nonlinear function of process inputs with adjustable time delays. Partial least squares (PLS) models predict a process output from a linear combination of a reduced set of independent variables called *latent variables* or *principle components*, which are themselves a linear combination of process inputs with adjustable time delays. The principal component analysis (PCA) eliminates correlations between process inputs. Both ANN and PLS are primarily steady-state methods reformulated for batch processes. Normally, the ANN and PLS process inputs and outputs are not incremental and do not have a time constant or process gain [11]. Model predictive control (MPC) provides a future trajectory (time response) for the change in a process output based on the linear combination of the effects of past changes in process inputs. The trajectory includes the effect of a steady state gain and time constant for self-regulating (exponential) responses or an integrating gain for integrating (ramping) responses, besides a time delay. Model identification software can be used to develop online dynamic estimators (ODE) with future trajectories for use in MPC [11].

ANN models predict nonlinear behavior, PLS and PCA models eliminate correlations between inputs, and MPC and ODE models provide the exponential and ramping time response for deviations about operating points.

Quality and Quantity of Data
To be capable of identifying a good quality model, the important inputs must have known changes that are at least five times larger than the noise associated with the input. Also, at least five changes in each input are needed. In other areas of process control, sufficient quantity and quality of data are ensured by conducting step or bump tests and pseudo random binary sequences (PRBS), in the case of MPC, or by designing experiments, in the case of PLS or ANN. It is rare that historical data can be completely relied on to generate these experimental models [3] [11].

To build good experimental models, there must be at least five changes in each important input that are each at least five times larger than the noise band.

Since each bioreactor batch may be worth hundreds of thousands to millions of dollars (for high value-added products), opportunities for introducing perturbations are severely restricted. Since each batch takes days to weeks to complete, even if perturbations could be made within the specification limits set by the plant's procedures for validating and managing change procedures, it might take months to years before enough data points were available to develop a MPC, PLS, or ANN model.

It takes months to years to get enough batches with
identifiable changes in process inputs that are large enough to
develop a MPC, PLS, or ANN model.

Processes that are subject to federal cGMPs, such as those that make
medicines, are required to define all critical process parameters
(parameters that could affect product quality) and the corresponding
"proven acceptable ranges." For fermentations that make
pharmaceuticals, parameters such as pH and temperature are typical
critical process parameters. In these cases, it would not be permitted to
deliberately perturb the pH or temperature outside the "proven
acceptable range" (which is a fairly narrow range of values for many
fermentations) just for the sake of obtaining model information. This is
because such perturbations would be defined as a process deviation and
could cause that lot to be rejected for the purposes of making product for
market [2] [8].

Golden Batch
In the early applications of PCA to batch processes, a *golden batch* for
multivariate statistical process control (MSPC) was defined to be the
median of the batch profiles for a fixed set of inputs. This definition is
misleading because the definition is not really representative of the best
possible theoretical or practical batch and it assumes fixed process inputs
that may not be achieved in plant operation. The best batch profile needs
to be found by theory and experimentation for any process input in order
to meet the control objectives. Chapter 8 on MSPC discusses the relative
merits of using multiway and model-based PCA to generate reference
batch profiles.

The "golden batch" is not the median of the batch profiles but
is the best batch profile achievable based on theory and
experimentation for a variable set of inputs.

Dynamic First-Principle Models
An opportunity that is now emerging is the use of dynamic first-principle
models in virtual plants. These models have accurate mass and energy
balances and kinetics, which include the activities of the nutrients and
cells in order to address many of the issues created by the lack of sufficient
test and analysis data [11]. These models can provide real-time inferential
measurements of broth composition without the noise and delay
associated with analytical results. The models can be run faster than real
time for rapid testing and for designing experiments. The results have
many uses, including discovering the best batch profile for the current set
of inputs, developing PCA logic for batch performance monitoring, and
identifying MPC, PLS, and ANN models. The models can fill in the
remaining profile of "in progress" batches for MSPC and predict key

performance indicators (KPI), such as batch end points, yield, and cycle times.

> *Dynamic first-principle models in a virtual plant can be used to develop KPI, "golden batch" profiles, batch performance monitoring logic, and MPC, PLS, and ANN models.*

Inferential measurements from even the best experimental and first-principle models do not eliminate the need for composition measurements. Field analyzers provide more frequent data with better timing than does laboratory analysis. This can reduce the model development and verification time from years to weeks. Initially, these measurements are needed to identify experimental and first-principle model parameters. Chapters 4 and 5 present techniques that show how these models can be used to provide a future (predicted) composition. The inferential measurement improves the reliability and reduces a process composition's noise and dead time for feedback control and optimization [11]. However, drift and unknowns are inevitable, and feedback correction is essential. A delayed and lagged composition value is synchronized with a field or laboratory measurement, and a portion of the difference is used for feedback correction of the future value [11]. A field analyzer or a probe, as simple as a dissolved carbon dioxide electrode, can provide a full correction within a batch phase. This contrasts with laboratory measurement, where multiple batches are generally needed to make a full correction.

> *Experimental and first-principle models do not diminish but instead accentuate the value of field analyzers, and vice versa.*

Batch, Fed-Batch, and Continuous Modes
Besides determining the availability and accuracy of measurements of the process inputs and outputs, the mode of operation also determines the selection, implementation, and performance of modeling and control methods. The main modes of operation are manual, sequential batch, fed-batch, and continuous. These modes affect how much variability is introduced into the process inputs, which in turn has an impact on process outputs and batch profiles.

In the manual mode, the operator starts and stops the addition of the process inputs. Typically, the concentration and duration of the amount added (batch charge) is specified, but the actual timing is inconsistent because of the differences in operator knowledge, skills, attentiveness, and goals. The batch record for pharmaceuticals enforces the timing as much as humanly possible, but this may not be the case for biofuel production, where manual enzyme and nutrient addition and yeast pitching are practiced.

Manual operation introduces variability into process inputs.

In the sequential-batch mode, a batch control system automatically prepares and sequences the addition of process inputs. Frequently, the charge is a totalized flow or a weight from a tank on load cells. However, the concentration of constituent raw materials and the prepared charge are seldom measured on line. In biopharmaceutical processes, a lab analysis is performed to confirm the composition of raw materials. However, lab procedures may not exist for measuring all the properties that affect a cell metabolism. In particular, not all the properties of corn are measured that affect the enzyme charges and batch cycle time for ethanol and the nutritional content and flowability of the byproduct of dry distiller's grain solids (DDGS) sold as animal and fish food. The variability of starch and nutrient content depending on field and weather conditions has a major impact on the performance of an ethanol plant. For the brewing industry, the quality of the water, such as alkalinity, affects the pH and taste of the product [12].

Sequential-batch operation reduces the variability of the mass and timing of process inputs but not necessarily the variability in the concentrations of charges.

Trace quantities of nutrients or contaminants may not be recognized as having an important effect on biological processes. The internal structure of the cells is not measured and is seldom even modeled. Consequently, the variability in the intended and undesired components of process inputs is not completely addressed by sequential-batch operation. Online analyzers and laboratory analysis typically focus on extracellular composition.

For processes that are subject to government cGMPS, such as in pharmaceutical manufacture, the raw material lots that the processes use must be assayed before the process's start to verify "fitness for use." In some cases, a guaranteed "certificate of analysis" from the supplier is sufficient. In general, scientists who are developing a manufacturing process must determine what parameters can affect product quality, which includes raw material concentrations and purities. After those parameters are identified the process must be tested for these parameters before it is implemented [8].

In the brewing, wine, and biofuel industries, awareness is growing that variability in the composition and quality of raw materials, such as hops, grapes, and corn, caused by location, weather, and water should be measured [12].

The largest suspect source of variability in well-automated batch and continuous operations are changes in the composition of inoculums for the pharmaceutical industry and raw materials and water for other industries.

In fed-batch operation, the flows of some of the process inputs are manipulated by a process controller. In these cases, variability in a process output is transferred to a process input. Common examples of fed-batch operation are the use of a pH controller to manipulate a reagent flow and a dissolved oxygen controller to manipulate an air or oxygen flow. Installing a glucose probe or using a first-principle model to inferentially measure glucose open up opportunities for a concentration controller to manipulate substrate feed. Fed-batch operation is sometimes called "semi-continuous" because control loops continuously manipulate process inputs. However, fed-batch is really still a batch operation because the discharge flow is zero until the end of the batch.

For continuous bioreactors, there is a continuous discharge flow from the bioreactor and continuous flows of process inputs. When the bioreactor is running in the production mode, there may also be some recycle of effluent to recover raw materials or to do further conversion. There may be a startup sequence and some of the same issues associated with batch sequences. Once the bioreactor reaches operating conditions, the focus shifts to throughput rate, load disturbance rejection, and process degradation as a result of contamination and recycle. Although continuous bioreactors can significantly increase capacity, practical issues have limited their use in industry primarily to treating activated sludge and producing biodiesel fuels. A few industrial biopharmaceutical processes are essentially "continuous," although they are better known as "perfusion" processes. These are characterized by continuous feeds and continuous withdrawal of broth (containing product). They run for several weeks or months per batch. However, they still eventually need to be stopped and cycled to start a new batch because of mutation and other concerns [8].

1-3.   Transfer of Variability

The variability of key process outputs, typically the composition and quality of products, is important in any plant. Theoretically, reducing product variability allows some plants to run at higher capacity or better efficiency by operating closer to constraints. However, there is a natural tendency not to run very close to product quality constraints in pharmaceutical batch processes, since a single excursion across such constraints could cause an entire lot to be deemed unacceptable for market, thereby wasting weeks of time and effort.

Another key factor in reducing variability is the number of runs needed in plant trials to evaluate suggestions for process modifications. For example, a typical bioprocess plant's coefficient of variation (COV), which is the "mean" divided by the "standard deviation," is about 10 percent, which is somewhat higher than the track record for small molecule chemical processes for pharmaceuticals. Most proposed plant modifications (e.g., nutrient changes, culture strain changes, etc.) have a suggested improvement of about 5 percent. The number of production fermentation runs needed to statistically verify that a 5 percent improvement is actually occurring, in an environment where plant COV is 10 percent, will be high. If plant variability could be improved such that COV is 5 percent, then significantly fewer expensive and time-consuming production fermentations would be needed to determine whether the suggested improvement was real or not. Nearly all proposed improvements are typically initially done at pilot scale. However, many changes do not scale well to production-size equipment (a common problem for bioprocesses), which means that production plant trials are also needed [8]. It is important to note that improvements do not automatically translate into dollars. Often a set point or mode of operation must be changed to realize the benefits [13].

For manual operation, operators' adjustments are a major contributor to product variability. To reduce this source of variability, batch automation software provides sequences that fix the timing and quantity of each process action. The remaining product variability can be attributed to changes in the cells and feeds and to the progression of the batch itself. Adding good feedback control, such as a well-tuned PID controller, controls a process output, such as pH or dissolved oxygen at a set point, by manipulating process inputs, such as reagent and gas flows. Changes that would have occurred in pH or dissolved oxygen now appear as changes in reagent and gas flow. Note that it is not possible to fix both the process outputs and process inputs and that attempts to fix all process inputs by discrete actions will not compensate for changes in the quantity, quality, and composition of process inputs. Feedback control offers plants the ability to compensate for unknowns by transferring variability from process outputs (controlled variables) to process inputs (manipulated variables). When the changes (disturbances) are measured, feedforward control can be added to feedback control to preemptively change a process input. This is achieved by modeling the relationship between the measured disturbance and the manipulated process input [3].

Modeling and process control open up new ways to minimize variability in a process's efficiency and capacity. Trying to predict and fix all the optimum steps in a profile of process inputs is infeasible. Instead, process control offers plants the opportunity to transfer variability in process outputs of direct or indirect value to process inputs and to provide a continuous profile that compensates for unknowns.

Discrete Process Actions
Often when manufacturing is faced with variability in process outputs, it chooses to focus completely on fixing process inputs. Operations and process design departments often request discrete process actions or steps to add specific amounts of reagents, gases, and substrate or they set cooling rates at specific times in the batch to anticipate batch requirements. "If ... then ..." is the predominant mode of thinking.

> *The predominant solution chosen by plants to reduce variability in batch operation is to set process inputs (perform discrete process actions) on a predictable time schedule.*

For continuous operation, there is theoretically a steady state, but even here unknown disturbances, such as influent composition and recycle, cause the process outputs to perpetually move. For batch operations, predicting input requirements is even more problematic in that there is no steady state, as evidenced by the batch profiles. The reagent demand, oxygen uptake rate, carbon dioxide evolution rate, and substrate consumption have a profile that is linked to the biomass and product profiles. The actual amount of a process input that is needed is a moving target in a batch, and feedback correction (i.e., closed loop control) is needed to compensate for unknowns in concentrations and biological mechanisms. Preprogrammed process actions (i.e., open loop control) may work well in laboratories and some pilot plants, and they can help a control loop get started (by initializing the controller outputs). However, preprogrammed process actions should not interfere with the proper operation of feedback controllers [14] [15]. Most of the conflicts between discrete actions and loops can be traced to a lack of fundamental understanding of the "what, how, and why" of basic feedback control for batch operations that have a moving target. For example, batch sequences may position a control valve in the remote output (ROUT) mode of a PID controller at various points in the batch in order to achieve a preprogrammed feed, coolant, or reagent profile. However, it would be better in this case to change the set point in a remote cascade (RCAS) mode and to allow the PID to correct for unknowns and disturbances through feedback control. This requires that configuration specialists understand the dynamics and controller tuning for batch profiles. To provide this understanding, chapters 2 and 3, on dynamics and tuning (respectively), discuss in detail the implications of the integrating response of batch processes compared to the self-regulating response of continuous processes.

Process actions can be useful to get a control loop started, but they should not retain too much control and interfere with the operation of basic feedback control.

Feedback Control

It is important that the feedback control be allowed to do its job. Not only is the immediate performance better, but over the long haul the performance will be more sustainable. The very definition of feedback control means that the control system is by nature self-correcting to some extent and thus able to deal with changing conditions. However, its effectiveness depends upon the tuning of the controllers. The section on adaptive control in chapter 3 shows how the controllers can be automatically tuned both initially and for changing final elements and batch conditions. The feedback control task is usually separated into two levels. The basic or lower level control may use proportional-integral-derivative (PID) controllers. To account for interactions and operating constraints and facilitate optimization, model predictive control (MPC) may be used for the advanced or upper level [3] [11]. In each level of feedback control, manipulated variables are adjusted to keep controlled variables at a set point or target.

Feedback control corrects for unknowns and unidentified disturbances.

Feedback control transfers variability in process outputs to process inputs so as to compensate for disturbances or differences in processing conditions [3]. For bioreactors, it is desirable to maximize this transfer so the variability in process outputs is minimized. For example, consider the following three situations: no feedback control, basic control (PID), and advanced control (MPC). Figure 1-3a shows a preprogrammed schedule of steps in the process inputs (reagent flow and oxygen-enriched air flow). The process outputs, pH substrate concentration, and dissolved oxygen (DO) vary with batch demand and with changes in the processing and concentrations of these and other process inputs. Also, the average values of process outputs are offset from the optimum because of bias errors in the measurements and unknowns in the kinetics and the process inputs. In figure 1-3b, PID loops manipulate the reagent and oxygen-enriched air flow to draw straight lines on pH and DO. Notice that there is still an offset from the optimum values of pH and DO and that the other process outputs, biomass and product concentration, still have variability and an offset compared to the best batch profile (golden batch) for the current set of inputs. In Figure 1-3c, an MPC whose set points are based on the golden batch profile manipulates the substrate and dissolved oxygen set points in order to provide a more uniform and optimum profile in the biomass and product concentrations [9]. Using an MPC has been demonstrated in two cases involving dissolved oxygen control to do a better job than a PID in dealing with noise and dead time. Chapter 5 on MPC details how the development of the virtual plant and an innovative translation of controlled variables have addressed most of the practical problems that have hindered biomass and product profile optimization to date. Figure 1-3c also shows how the pH may be trimmed to account for pH drift or for the pH requirements during different phases in the fermenter, such as the saccharification, growth, and product formation phases for ethanol production.

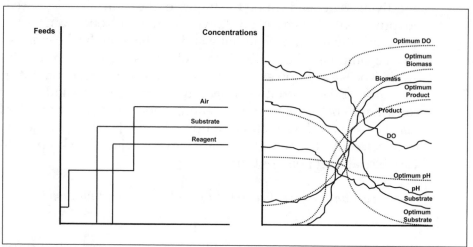

Figure 1-3a. Bioreactor Inputs and Outputs for No Feedback Control

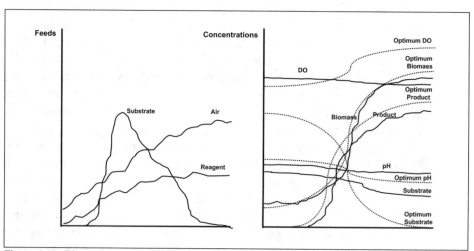

Figure 1-3b. Bioreactor Inputs and Outputs for Basic PID Control

In summary, good batch manager software is essential. When it is combined with good measurements and final elements, manufacturers can more effectively track down the sources of variability and implement

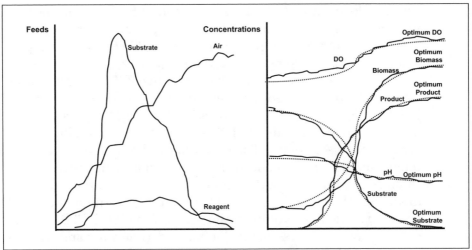

Figure 1-3c. Bioreactor Inputs and Outputs for Advanced MPC Control

varying levels of process control so as to transfer variability from process outputs to process inputs. Transferring variability is the key to identifying abnormal batches and achieving more optimum and consistent operating points, profiles, and end points.

Selecting MSPC, PLS, and ANN Inputs

The transfer of variability profoundly effects which inputs and outputs are appropriate for MSPC batch analysis and for PLS and ANN models, yet this is a topic that is not usually discussed in literature or recognized by the user. For pH control, the input for MSPC, PLS, and ANN would change from pH to reagent flow once the loop is put in the automatic mode (i.e., the loop is closed).

> *Feedback control transfers variability from process outputs to process inputs and profoundly affects which inputs are chosen for MSPC batch analysis and PLS and ANN models. As loops are closed, the MSPC, PLS, and ANN inputs move from controlled variables to manipulated variables.*

Setting Up Feedback Loops

How the loops for a bioreactor are set up will depend on the product and the degree of sophistication needed to meet market requirements. However, it is useful to consider how the different types of cell cultures affect the selection of the process outputs to be controlled and the process inputs to be manipulated. For example, figure 1-3d shows the setup of the basic feedback control loops for a fungi cell culture used for antibiotic production. The temperature controller manipulates jacket temperature, the pH controller manipulates ammonia reagent flow, the dissolved oxygen controller output is split to manipulate air flow and agitation based on load, and finally the glucose concentration controller manipulates glucose feed rate. If these loops have fixed set points, the MSPC, PLS, and ANN inputs would be the PID controller outputs (i.e., manipulated jacket temperature, ammonia flow, air flow, pressure, agitation, and glucose flow). If the set points of these control loops were continually manipulated for more optimal operation by an MPC, then the MSPC, PLS, and ANN inputs would be the MPC outputs (i.e., manipulated temperature, pH, dissolved oxygen, and glucose concentration set points). The seed culture charge concentration and timing and the off-gas concentrations of components, such as oxygen and carbon dioxide, are important process inputs and outputs for analysis and modeling. They can also be used to infer the oxygen uptake rate (OUR) and carbon dioxide evolution rate (CER). The MSPC, PLS, and ANN outputs nearly always include biomass concentration and/or growth rate and product formation rate and/or concentration since these are essential to a bioreactor's operation.

An increase in head pressure increases the equilibrium concentration for dissolved oxygen via Henry's Law and hence increases the driving force for mass transfer. However, pressure may not be manipulated for dissolved oxygen because it also increases the concentration of dissolved carbon dioxide, which is usually not measured.

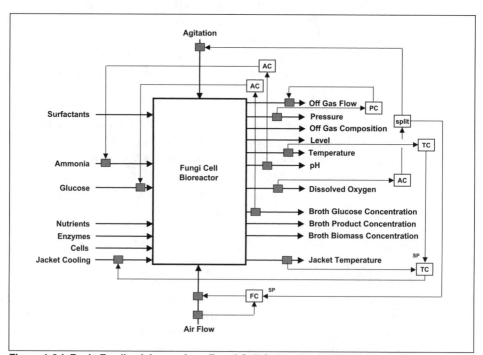

Figure 1-3d. Basic Feedback Loops for a Fungi Cell Culture for Antibiotic Production

Figure 1-3e shows the setup of the basic feedback control loops for a simultaneous saccharification and fermentation of a yeast cell culture used to produce ethanol from corn. The loops are similar to figure 1-3d except that the reagents are different and there is no dissolved oxygen control. The substrate is a starch or cellulose that is converted into glucose (saccharification) by an enzyme and then fermented to ethanol by yeast. In this case, the glucose concentration controller manipulates the liquefied starch or cellulose feed. This feed pH is controlled by adding ammonia and sulfuric acid. The nitrogen, phosphorous, minerals, and trace elements in the feed are balanced by adding various chlorides, phosphates, sulfates, and yeast extract. In the bioreactor, the yeast, ethanol, and enzyme concentrations are important. Once the loops are closed, the MSPC, PLS, and ANN inputs become jacket temperature, ammonia and sulfuric reagents, and starch or cellulose feed. The amounts and timing of the enzyme and yeast charge (process inputs) and the off-gas carbon dioxide concentration (process output) would be important inputs for analysis/modeling.

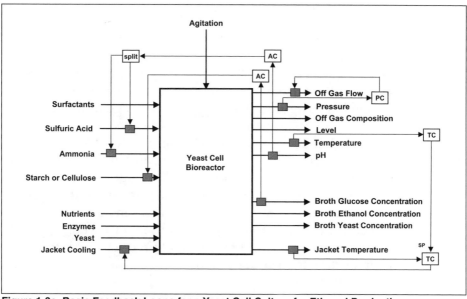

Figure 1-3e. Basic Feedback Loops for a Yeast Cell Culture for Ethanol Production

Figure 1-3f shows the setup of the basic feedback control loops for a bacterial cell culture used in biopharmaceutical production. The setup of the loops is similar to figure 1-3d except that the air flow may be enriched by oxygen. Acid does not need to be added for bacterial cell culture fermentations unless, perhaps, as a contingency to counter the mistaken addition of too much base. Bacterial cultures generate two sources of acid naturally: CO_2 evolution (carbonic acid), and acetate formation (acetic acid).

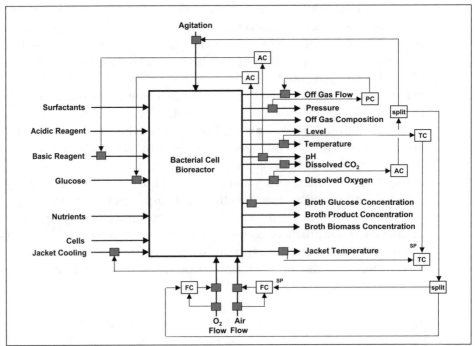

Figure 1-3f. Basic Feedback Loops for a Bacterial Cell Culture for Pharmaceutical Production

Figure 1-3g shows the setup of the basic feedback control loops for an animal cell culture used in biopharmaceutical production. The setup of loops is similar to figure 1-3f except that the reagents are carbon dioxide and sodium carbonate or sodium hydroxide, there are two substrates (i.e., glucose and glutamine), and actual serum and/or various amino acids, growth factors, vitamins, and minerals are added to duplicate a blood serum as closely as possible. Also, vessel pressure and agitation are not manipulated. An increase in vessel pressure increases the dissolved carbon dioxide concentration, and an increase in agitation breaks the animal cells, which are more fragile because they do not have cell walls. Besides the usual biomass, substrate, and product concentrations, it may also be important to measure byproducts, such as lactic acid, which indicate cell health.

1-4.   Online Indication of Performance

A key performance indicator of a bioreactor's process efficiency is yield. It can be computed from the product mass at the end of the batch, which is the product end point concentration (mg/liter). Though yield is commonly stated just in terms of mg/liter of product, the fundamental definition of *yield* takes into account the batch volume and substrate charged. In this book, the yield is expressed in terms of a particular substrate and is computed as the kg/liter of product multiplied by the

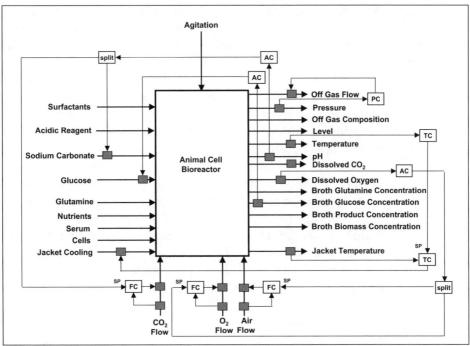

Figure 1-3g. Basic Feedback Loops for an Animal Cell Culture for Pharmaceutical Production

broth volume (liters) and divided by the total mass of substrate charged (kg). This dimensionless number for yield can be compared to the yield coefficients used in the kinetic expressions. If the actual yield is divided by the yield coefficient and multiplied by 100 percent, then the relative yield becomes a percentage of maximum practical yield. This substrate yield is important because of the large quantities of substrate used. For animal cell cultures, the serum can be the largest material cost.

A key performance indicator of a bioreactor's process capacity is production rate. It can be computed as the product mass at the end of the batch divided by the batch cycle time (kg product/hour). The time spent holding the batch or waiting on other unit operations is important for plant capacity. However, an increase in the bioreactor capacity eventually shows up in additional plant capacity as these bottlenecks are identified and eliminated. Greater bioreactor capacity means more time for maintenance and process improvements. If the plant is not sold out, it also means more time that the plant can be idle. Higher product concentrations can also translate into reduced purification and waste treatment costs.

Greater bioreactor production capacity can result in improved maintenance and processes in addition to lower downstream processing costs.

Expressing bioreactor yield and production rate as a percentage of the maximum practical value demonstrated in industrial bench-top and pilot plant systems provides a benchmark and better scaling for recognizing changes in bioprocess efficiency and capacity. The prediction of yield and production rate is filtered to reduce noise and displayed on trend charts in the operations control room. Online trending of properly processed key performance indicators improves the performance of operations, control system support, maintenance, and process support by increasing the plant's awareness of where it has been and where it is going. In figure 1-4a, a simple linear program (LP) optimizer can find an MPC's optimum operating point, which is the vertex intersection of operating limits for the manipulated variables (i.e., DO and glucose) and the controlled variables (i.e., biomass and product concentration). The optimum vertex depends on the value of yield and capacity with changes in market demand, operating costs, and process limits. If the vertex moves, a high-fidelity first-principle model and real-time optimization are needed to find the new vertex.

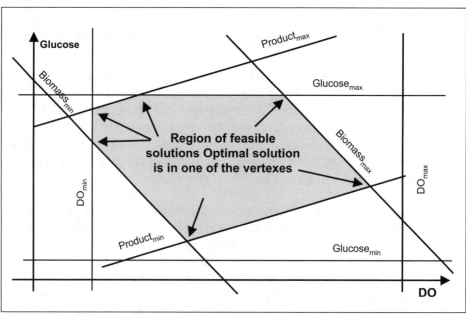

Figure 1-4a. Variation in Optimum DO and Glucose

Online trending of key performance indicators improves the technical support, operation, and maintenance of the bioreactor and its control systems.

Virtual plants with dynamic first-principle models can predict key performance indicators and process variables as the batch progresses. They do this by running faster than real time so as to create profiles into the future of yield, capacity, biomass, and product. Additionally, these virtual plants running real time can be automatically adapted by a combination of model predictive control and neural networks so as to provide fast and reliable inferential measurements that become the controlled and optimization variables. Figure 1-4b shows this general setup, which we discuss in chapter 5 on the virtual plant.

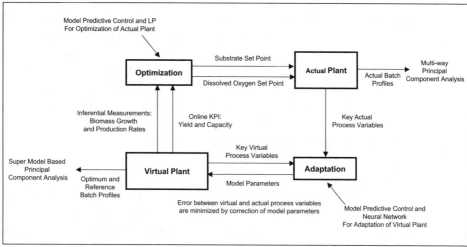

Figure 1-4b. Online KPI and Inferential Measurements from an Adapted Virtual Plant

> *An adapted virtual plant can provide online predictions of yield and capacity as well as inferential measurements of growth and production rates for optimization.*

The virtual plant can also act as a knowledge base for exploring the opportunities for making additional measurements. The virtual plant can quantify and demonstrate the value of additional Coriolis meters for mass flow, density, and viscosity measurements as well as additional conventional and analytical measurements. For example, in ethanol production a direct density and inferential viscosity measurement of the laminar fermenter feed from a Coriolis meter and differential pressure sensor may provide important information about starch and dextrin content. This can be useful for optimizing the enzyme charge. Similarly, a pressure sensor on the bottom of the fermenter may provide an accurate inferential measurement of the loss-in-weight of the carbon dioxide evolution rate, which is useful for adapting the virtual plant's kinetics.

1-5.   Optimizing Performance

Inferential measurements of biomass, product, and substrate concentrations can be used as the controlled variables for model predictive control. The set points or targets for these concentrations are the golden batch profiles used to achieve production objectives. Alternately, inferential biomass growth rate and product formation rates can be used as the controlled variables to help remove the effects of bias errors. They can also be used to prevent the MPC from attempting to force growth for formation rates that are too fast or too slow while the MPC is attempting to match a profile. In this case, the set points or targets are the slopes of the golden batch profiles. In both cases, the penalties on error for the controlled variables for the MPC and an LP optimizer can set the relative importance of the maximization of yield versus capacity.

The use of biomass growth and product formation rates as MPC-controlled variables eliminates bias errors and avoids the risk of forcing rates that are too fast or slow.

1-6.   Process Analytical Technology (PAT)

Origins of PAT
In the 1950s, a new citric acid production facility had 50,000 L fermenters. Batches ran for a month, and the substrate was molasses from sugar cane. Forty years later the same plant produced a higher-quality citric acid using 150,000 L fermenters. Batches were completed in less than a week, and expensive cane sugar substrate was replaced by fructose from corn starch.

The Food and Drug Administration (FDA) has long held that quality cannot be tested into products. Quality must be built into a process. Traditionally, quality has been verified after the manufacture of a product. Process steps that are successfully operated in a repeatable manner are accepted as being validated. Generally, a process step is considered validated after three consecutive successful batches.

This methodology provides a high probability that a high-quality product will be attained, but the uncertainty of managing process changes through regulatory submissions has discouraged innovation. The reason behind the Process Analytical Technology (PAT) initiative is to allow pharmaceutical manufacturers to continuously improve processes along with the rest of the chemical process industry and at the same time assure a high-quality product.

FDA Initiative
In an effort to increase the safety, efficiency, and affordability of medicines, the FDA has proposed a new framework for regulating pharmaceutical development, manufacturing, and quality assurance. The primary focus of the initiative is to reduce variability by gaining a better understanding of a process than can be obtained by the traditional

approach. Note that PAT is a "guidance." PAT is not contained in the Code of Federal Regulations, and companies are not required by law to use PAT.

A draft report explaining the new regulatory framework was published in 2002 and titled [16]:

Pharmaceutical CGMPs for the 21st Century – A Risk-Based Approach.

Along with the final version of this report, the "Guidance for Industry" was published in September 2004. It was titled [17]:

PAT – A Framework for Innovative Pharmaceutical Development, Manufacturing, and Quality Assurance.

Although *PAT* stands for "Process Analytical Technology," it has become an acronym for designing quality into a process. PAT is formally defined in section IV of the "Guidance for Industry":

"The Agency considers PAT to be a system for designing, analyzing, and controlling manufacturing through timely measurements (i.e., during processing) of critical quality and performance attributes of raw and in-process materials and processes, with the goal of ensuring final product quality. It is important to note that the term analytical in PAT is viewed broadly to include chemical, physical, microbiological, mathematical, and risk analysis conducted in an integrated manner. The goal of PAT is to enhance understanding and control the manufacturing process, which is consistent with our current drug quality system: quality cannot be tested into products; it should be built-in or should **be** by design."

The focus of this new regulatory environment is on discovering process variation and controlling that variation when it might contribute to patient risk. The process variation is discovered by identifying and measuring critical quality attributes in a timely fashion. In this way, processes can be developed and controlled in such a way that the quality of product is guaranteed.

Being able to analyze the sources of process variability and control that variability could provide engineers with a basis for making equipment changes and scaling up processes.

Business Drivers
A biopharmaceutical manufacturer that can develop a process rapidly and efficiently, according to Pisano [18], reaps the benefits of accelerated time-to-market; rapid ramp-up of yield, quality, and capacity from initial to normal long-term levels; and a stronger proprietary position. Although

Pisano's comment was made in the context of the then-existing regulatory environment, the goal of the PAT framework is to help manufacturers reap just these benefits. The "Guidance for Industry" states:

"Gains in quality, safety and/or efficiency...are likely to come from:

- Reducing production cycle times by using on-, in-, and/or at-line measurements and controls

- Preventing rejects, scrap, and re-processing

- Real-time release

- Increasing automation to improve operator safety and reduce human errors

- Improving energy and material use and increasing capacity

- Facilitating continuous processing to improve efficiency and manage variability. For example, use of dedicated small-scale equipment (to eliminate certain scale-up issues)."

PAT Tools

The "Guidance for Industry" defines PAT tools and categorizes them as:

- "Multivariate tools for design, data acquisition and analysis

- Process analyzers

- Process control tools

- Continuous improvement and knowledge management tools."

PAT opens up opportunities for analyzers and for the integrated use of tools for analysis, control, and modeling to improve plant performance.

Process control as well as continuous improvement and knowledge management are the two PAT tool categories that are most applicable to the subject of this book: the study of bioreactor performance.

One type of process control is the optimization of key process indicators (KPI). Examples of KPI discussed in this book are the "predicted product yield" and the "predicted production rate" of bioreactors. Basic feedback control loops are improved and then dynamic first-principle models are integrated with model predictive control in a virtual plant to ensure product quality while maximizing these KPI.

References

1-1.   McMillan, Gregory K., *Biochemical Measurement and Control*. Reprint via ProQuest UMI "Books on Demand", ISA, 1987.

1-2.   Alford, Joseph S., "Bioprocess Control: Advances and Challenges." Presentation at CPC-7 to be published in *Computers and Chemical Engineering*, 2007.

1-3.   Blevins, Terrence L., McMillan, Gregory K., Wojsznis, Willy K., and Brown, Michael W., *Advanced Control Unleashed: Plant Performance Management for Optimum Benefits*. ISA, 2003.

1-4.   Aiba, Shuichi, Humphrey, Arthur E., and Millis, Nancy F., *Biochemical Engineering*. 2d ed. Academic Press, 1973.

1-5.   *Chemical Engineering*. "CE Refresher – Liquid Agitation." Reprint from *Chemical Engineering*, 1985. McGraw-Hill.

1-6.   Nienow, Alvin W., Langheinreich, Christian, Stevenson, Neil C., Emery, Nicholas A., Clayton, Timothy M., and Slater, Nigel K. H., "Homogenisation and Oxygen Transfer Rates in Large Agitated and Sparged Animal Cell Bioreactors: Some Implications for Growth and Production." *Cytotechnology* 22 (1996): 87-94. Kluwer Academic Publishers.

1-7.   McMillan, Gregory K., and Cameron, Robert, *Advanced pH Measurement and Control*. 3d ed. ISA, 2005.

1-8.   Alford, Joseph S., Email correspondence, 2006.

1-9.   Grant, Wilson, McMillan, Gregory K., and Boudreau, Michael, "PAT Benefits from the Application of Modeling and Advanced Control of Bioreactors." *Emerson Exchange*, October 2005.

1-10.   Alford, Joseph S., "Bioprocess Mass Spectrometry: A PAT Application," *The Journal of Process Analytical Technology*, May-June 2006.

1-11.   Trevathan, Vernon L. (editor), "Process Modeling" (chapter 12) and "Advanced Process Control" (chapter 13). In *A Guide to the Automation Body of Knowledge*. ISA, 2005.

1-12.   Anderson, Dave, "Watch the Water." *Intech*, November 2005.

1-13.   McMillan Gregory K., and Weiner, Stan, "The Bad Hall of Fame" (Control Talk). *Control*, March 2006.

1-14.   McMillan, Gregory K., and Weiner, Stan, "Life Is a Batch" (Control Talk). *Control*, June 2005.

1-15.   McMillan, Gregory K., and Weiner, Stan, "Grounded in Reality" (Control Talk). *Control*, December 2005.

1-16.   "Pharmaceutical CGMPs for the 21st Century–A Risk-Based Approach." Final Report, Department of Health and Human Services, U.S. Food and Drug Administration, September 2004.

1-17.   "Guidance for Industry. PAT – A Framework for Innovative Pharmaceutical Development, Manufacturing, and Quality Assurance." Department of Health and Human Services, U.S. Food and Drug Administration, September 2004.

1-18.   Pisano, Gary P., *The Development Factory: Unlocking the Potential of Process Innovation*. Harvard Business School Press, 1997.

Chapter 2:
Process Dynamics

Chapter 2

Process Dynamics

2-1. Introduction

The dynamics of a process response can be modeled by three types of parameters: dead time, time constant, and process gain. The dynamics are identified from the plant's operation, and they determine both the process's ultimate performance and the actual performance of the control system. The process dynamics are used to compute PID controller settings. They could also be employed directly to predict the trajectory of process outputs from past process inputs in future applications of model predictive control (MPC) for bioreactors. Although MPC technology has been used almost exclusively on continuous processes, recent batch applications are promising, and it is anticipated that bioreactor applications are possible.

Section 2-2 of this chapter discusses the largest sources of dead time in the automation and process systems, the impact of total dead time on the controller's capability and tuning, and the procedure for estimating the performance based on the tuning settings. Section 2-3 details the characteristics of a self-regulating process response, the way small time constants effectively become dead time, and the reasons why the location of the largest time constant affects control. Section 2-4 delves into the integrating process response of the more important process variables as well as the implications of this response for process testing and controller tuning,

Learning Objectives

A.   Understand how dead time sets the ultimate performance limit and maximum possible aggressiveness of the controller tuning.

B.   Learn how the design of equipment and automation systems affect dead time.

C.   Become familiar with ways to estimate loop performance from tuning settings.

D.   Be aware of how limit cycles originate from the final element resolution.

E.   Learn the three basic parameters used to define the process response.

F.   Recognize the differences between self-regulating and integrating responses.

G.   Be able to convert self-regulating response parameters into an integrating parameter.

2-2.   Performance Limits

Impact of Loop Dead Time

Control loop dead time is the period of time after a change in the controller output (manipulated variable) in which there is no recognizable change in the process variable (controlled variable). For feedback control, the controller output must change a process input in order to cause a change in a process output that is measured and seen by the controller. The performance of the control loop is ultimately limited by the loop dead time, which is the time it takes to make and see a change. However, the actual performance is determined by the controller tuning. The dead time sets how aggressively a controller can be tuned without becoming unstable. In practice, there is always a tradeoff between performance and robustness (that is, the controller's relative ability to be stable despite changes in process dynamics). In the process industry, controllers are generally tuned with settings that are far from the performance limit allowed by process dynamics [7]. This is particularly true for the primary loops on bioreactors because the process variable's rate of change as determined by a large process time constant or a small integrating process gain is slow compared to the size of the process dead time.

The ultimate limit to controller performance is determined by the loop dead time, but the actual limit is usually set by the controller tuning.

The control loop must respond to changes in its set point and loads (disturbances). The disturbances can originate from changes to any of the process inputs, as discussed in chapter 1, such as flows, charges, and intracellular metabolism and composition. Set point changes to primary loops (DO, pH, and temperature targets) are rather minimal, although the optimum pH and temperature may change from the growth to production phases. However, the set points to the secondary loops (changes in gas or reagent flow, agitation, and coolant demand) are continual and cover a wide range. This is so they can deal with the one to three order-of-magnitude changes in biomass during the first half of the batch [1].

As MPC or real-time optimization (RTO) applications are developed, set point changes to the primary loops become more frequent. Fortunately, the bioreactor response to disturbances is slow because the volume is large compared to flow rates and the growth, product formation, and mutation rates are slow.

Sliding stem control valves are rarely used within the sterile envelope of the bioreactor. Here, the *sterile envelope* is defined (envisioned) as including the bioreactor and the piping network inside the sterile filters. The few control valves that are used here are easy to clean and sterilize [1]. One type available is a clean-in-place (CIP), sanitize-in-place (SIP), and self-draining sliding stem angle control valve with a packless design, metal diaphragm, and a low-friction roller-bearing mechanism [2]. A wider spectrum of valves is used outside of the sterile envelope. A *sterile filter* is often used downstream of the valve to ensure that whatever is going into the fermenter is sterile.

The control valve's resolution and dead band should be determined per ISA standards ISA-TR75.25.01-2000 (R2006) and ISA-TR75.25.02-2000 (R2006) on valve-response testing and measurement. Although it is difficult to generalize, sliding stem valves that have a packless or low-friction design with digital positioners have a resolution limit of 0.05 percent to 0.1 percent and a dead band of 0.1 percent to 0.2 percent. Rotary control valves tend to have a larger resolution limit and dead band. Control valves without positioners have an order-of-magnitude or larger resolution limit and dead band [3] [4] [5].

For the small actuators used on valves for bioreactor control, the prestroke dead time and stroking lag times associated with pressure changes in the actuator are negligible. The largest source of dead time in a valve response is the resolution limit and dead band whenever the rate of change is slow in the controller output. The dead time can be estimated as the resolution limit or dead band for changes in the same direction and opposite direction, respectively, divided by the rate of change of the controller output [3] [4] [5]. However, as the total size of the change approaches the resolution limit, the dead time gets much larger, particularly in pneumatic positioners [5]. Some digital positioners can be tuned to minimize this increase in dead time. The larger and faster changes in the controller's output from less sluggish tuning settings can minimize this dead time.

The control valve dead time from resolution and dead band can be minimized by using a digital positioner and faster controller tuning settings.

The start of a change in flow as a result of a change in speed is almost instantaneous unless a dead band has been configured in the variable speed drive (VSD) to prevent a response to noise. The rate of change of speed might be limited in the VSD for motor load, but the rate limiting is usually fast enough so that it is not a consideration.

In smaller-scale bench-top and pilot plant–size bioreactors, there is significant use of peristaltic pumps for liquid additions such as base and substrate. This is because it is easy to take the tubing out of the pump roller bar mechanisms and clean and sterilize the tubing associated with these feeds [1].

Impact of Controller Tuning Settings

In a single-loop configuration, the output of the process controller (e.g., DO, pH, or substrate) goes directly to a VSD or a control valve. In a cascade loop, the output of the primary process controller becomes the set point of a secondary controller. The most common secondary controller is the flow or speed controller. These secondary loops compensate for load changes and disturbances before they affect the primary process controller. However, the secondary controller response must be five times faster than the primary controller response, or interaction will develop between the two controllers. The secondary controller must be tuned for a fast response. For flow and speed control, this corresponds to a reset time of less than six seconds. The performance of a cascade control system that has a slow secondary loop may be worse than if the primary controller output would set the VSD's variable frequency directly without tachometer feedback.

The secondary controller must be tuned to be five times faster than the primary controller or cascade control can do more harm than good.

According to the ISA and Fieldbus standards, the tuning settings for a proportional-integral-derivative (PID) controller are gain (K_c), reset time (T_i), and rate time (T_d). The gain setting is dimensionless, and the time settings are in seconds per repeat for reset and in seconds for rate. For reset time, the setting may be listed in terms of seconds with the seconds per repeat implied. In some controllers, proportional band in percent ($100\%/ K_c$) is used instead of gain, and reset action in repeats per minute ($1/ (T_i/60)$) is used instead of a reset time in seconds per repeat. Since these settings are inversely related to the ISA settings, it is critical that users understand what type of tuning settings is used as well as their units. In most cases, derivative action is not used (rate time is zero), which results in a proportional-integral (PI) controller. The effect of PID form and structure on performance is discussed in chapter 3, on basic feedback control.

Another important consideration is that users set the controller action so it is the opposite of the process's action if any reverse action of a final element is accounted for by a signal reversal. This reversal can be achieved in field devices, but it is preferable that it be done for maintainability and visibility in the blocks in the configuration, such as the "splitter" and the "analog output" blocks. Barring any improper signal reversals, the controller action is "reverse" if the process action is "direct" and vice versa for a single manipulated variable. For a "reverse" action controller, an increase in the process variable above set point decreases the controller output. A reverse-acting (fail-open or increase-to-close) final element is usually not a concern except perhaps for a coolant or vent control valve.

Signal reversals typically occur when the controller output sets multiple flows (i.e., multiple manipulated variables) by either split-ranged or simultaneous action. The signal reversal is normally done in "splitter" and "analog output" (AO) blocks for split-ranged and simultaneous action, respectively. The output action can also be reversed in the controller block, but this practice is not recommended unless the PID is directly addressed and there is no "splitter" and AO block.

Most industrial fermentations only use a base addition since the metabolic products are acidic in nature. However, in some applications a pH controller is split-ranged between acid and base flows where the signal is reversed for the acid flow. For example, the acid flow decreases, and when this flow reaches zero, the base flow increases as the controller output increases, which increases the pH (direct process action). In this case, the pH controller has reverse control action, and the "splitter" block reverses the signal for the acid flow in its output array specification.

Although a DO controller typically manipulates one process input (a single flow or agitation speed), in some applications a DO controller is set up to simultaneously manipulate air and oxygen flows in order to provide an oxygen-enriched gas flow. For example, the air flow decreases and the oxygen flow increases as the controller output increases. In this case, the DO controller has reverse action, and the analog output block for the air flow has a signal reversal (i.e., increase to close) specified.

If the conversion between gain and proportional band and between reset action and reset time or controller action is wrong, nothing else matters.

Equation 2-2a for integrated error is derived, as shown in appendix D, from the response of a standard form of the PI controller to a load upset where the open loop error (E_o) is the magnitude of the error if the controller is in manual. The integrated error (E_i) is the total area between the process variable and the set point for the controller in automatic [3] [4]. The equation shows that the integrated error is inversely related to controller gain and directly proportional to reset time.

$$E_i = \frac{1}{K_o * K_c} * T_i * E_o \qquad (2\text{-}2a)$$

where:

E_i =   integrated error (% seconds)
E_o =   open loop error (%)
K_c =   controller gain
K_o =   open loop gain (also known as process gain) (%/%)
T_i =   controller reset time (seconds)

If we take the ratio of the integrated errors at a constant open loop gain for new and original tuning settings and subtract this ratio from 1, the open loop error cancels out. We then end up with the expression within the parentheses in equation 2-2b, which gives an estimate of the fractional improvement in integrated error for load changes achieved by more aggressive tuning. If we multiply this fraction improvement by the control error, the result is the possible improvement in the standard deviation as a result of better tuning. The equation assumes that the loop is stable. If the loop is oscillating, it might be better to have less aggressive tuning, but this case is rare except at the start of the batch when demand is extremely low. Equation 2-2b does not address the portion of the standard deviation that is caused by limit cycles or noise since these are set by the devices, equipment, and process. Though equation 2-2a was derived for load changes, it also gives a relative indication of the performance improvement for set point changes. For cascade control, equation 2-2a should first be used to improve the tuning of the secondary loop before it is applied to the primary loop.

$$\Delta S_c \cong \text{Maximum} \left[\left(1.0 - \frac{K_{co}}{K_{cn}} * \frac{T_{in}}{T_{io}} \right), 0.0 \right] * \Delta E_c \qquad (2\text{-}2b)$$

where:

ΔE_c =   control error caused by load or set point changes (%)
K_{cn} =   new controller gain
K_{co} =   old controller gain
ΔS_c =   improvement in standard deviation from new controller tuning (%)
T_{in} =   new controller reset time (seconds)
T_{io} =   old controller reset time (seconds)

The control error (ΔE_c) that is caused by load or set point changes can be estimated from the mean absolute error (MAE_c) over a test period. The current test time is the module's execution time added to the last value of the test time multiplied by a fractional "forgetting factor" determined by the supervision time, as detailed in equation 2-2c. Similarly, the integrated absolute error is the absolute error added to a last value of the integrated absolute error forgotten to a degree again determined by the supervision time, as detailed in equation 2-2d. The mean absolute error is just the integrated absolute error divided by the test time, as shown in equation 2-2e. Finally, the control error (ΔE_c) for use in equation 2-2b is computed, per equation 2-2f, as the absolute difference between the mean absolute error and the standard deviation of the capability of the process (S_{cap}) set by noise. This can be estimated from the moving rate average of successive differences of the deviation from set point, as illustrated by equations 2-2g and 2-2h. The supervision time is typically either the duration of an operator shift or a batch phase. If the configuration is not set up so the set point tracks the process variable when the PID control algorithm is in the manual or remote output modes, then the error should be zeroed out until the controller is actually in service (automatic, cascade, or remote cascade modes).

$$T_n = T_e + (1 - 1 / T_s) * T_{n-1} \tag{2-2c}$$

$$IAE_n \cong \ | \ CV_n - SP_n \ | \ + (1 - 1 / T_s) * IAE_{n-1} \tag{2-2d}$$

$$MAE_c \cong IAE_c / T_n \tag{2-2e}$$

$$\Delta E_c \cong \ | \ MAE_c - S_{cap} \ | \tag{2-2f}$$

$$MR_n = r * [\ | \ CV_n - SP_n \ | \ - \ | \ CV_{n-1} - SP_{n-1} \ | \ ] + (1 - r) * MR_{n-1} \tag{2-2g}$$

$$S_{cap} \cong MR_n / 1.128 \tag{2-2h}$$

where:

CV_n	=	controlled variable for current scan (%)
CV_{n-1}	=	controlled variable for last scan (%)
IAE_n	=	integrated absolute error for test time (% sec)
MAE_n	=	mean absolute error for test time (%)
MR_n	=	moving rate average at scan n (%)
r	=	weighting factor (0.02)
SP_n	=	set point for current scan (%)
SP_{n-1}	=	set point for last scan (%)
S_{cap}	=	standard deviation of the capability of process as determined by noise (%)
T_e	=	module execution time (seconds)
T_s	=	process supervision time (seconds)
T_n	=	current test time (seconds)

An adaptive controller has been developed that identifies the process model and the new controller tuning settings for equation 2-2b. This makes it possible to estimate the possible improvement in the standard deviation through better tuning [5] [6] [7].

The improvement in control loop performance can be estimated from the change in tuning settings and the mean absolute error.

The controller tuning's dependence on process dynamics, including dead time, is detailed in chapter 3. It is important to remember that while the most aggressive tuning settings and the ultimate performance depends on the relative size of the dead time, the actual performance still depends on the tuning settings used. In other words, if a controller is detuned, it will perform the same as a controller with more dead time.

Sources of Dead Time

The purpose of a control system is to recognize and deal with change. Figure 2-2a shows the path from a change in controller output to a change in the process variable seen by a DO, pH, and substrate controller. A dead time or a time lag is associated with each block in the path. When DO is controlled by manipulating agitation speed, there are no piping or injection time delays. The total loop dead time can be estimated as the sum of the time delays and time lags of each block [3] [4] [5]. It can be visualized as the total delay required for a signal to make one cycle around the loop. The starting point of the change can be anywhere in the loop.

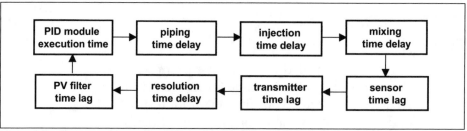

Figure 2-2a. Sources of Loop Dead Times for DO, pH, and Substrate Control

The controller has a dead time that is, on average, one-half of the module execution time, which is typically one to five seconds. Excluding resolution limits, the variable speed drive's dead time is negligible unless a dead band has been added for noise rejection.

Once the flow is changed, it takes some time for the change to get into the broth. The injection delay is the time it takes for the change in flow to make it into the broth. It includes any transportation delay in the piping from the point of the mixing of two streams, such as might occur in oxygen enrichment for DO control. If there is a dip tube, a transportation delay exists in the tube if the stream composition changes from mixing of streams or backfilling of the dip tube with broth when the pump is off. The transportation delay is equal to the piping or dip tube volume divided by the flow. Since reagent flows are so small, using a dip tube creates a huge reagent delivery delay for pH control [6]. Fortunately, the reagent is normally dumped on the surface. However, if the agitation is not noticeably breaking the surface or there is more radial (swirling) than axial (vertical) movement of the broth, there is still a considerable delay before the reagent is pulled down from the surface into the broth [6]. If there is more than one reagent, each reagent should have its own piping and dip tube. Separate reagent piping prevents the transportation delay detailed in the first expression of equation 2-2i. This delay is the time for the propagation of a change in reagent concentration from the point of mixing of reagents to the point of injection.

Once the change in flow has entered the broth, it takes time for it to be sufficiently dispersed and transported to the sensor. The location of the sensor relative to the location of the injection is a key factor. If the sensor is too close to the injection, short-circuiting occurs, producing an inconsistent reading. If the sensor is too close to the wall or a baffle, the lower fluid velocity translates into a slower dispersion and a slower inherent electrode response, which is greatly aggravated by an increased propensity for fouling [6]. In the extreme case of stagnation, the sensor does not see a broth that is representative of the mixture. If the sensor was located externally, there would also be a transportation delay associated with the piping from the bioreactor to the sensor.

Slow fluid velocities increase the fouling and time lag of electrodes.

The process dead time caused by liquid mixing (ignoring transportation, injection, stagnation, and sensor delays) for a well-mixed vessel with axial agitation can be estimated as the turnover time. This is approximately the broth volume divided by the volumetric injection flow and the agitator pumping rate [6]. Some bioreactors are not well mixed, so the actual dead time resulting from mixing is larger than the turnover time. This is especially the case for animal cell cultures, where agitation is minimal since the cells are easily damaged due to lack of a cell membrane. The largest dead time for a pH or substrate sensor inserted into a bioreactor is usually the sum of the injection transportation delay and the mixing time delay. This process dead time can be approximated by equation 2-2i for well-mixed bacterial, fungi, and yeast cultures and is rivaled only by the dead time from the final element's resolution limit.

If the agitation is not breaking the surface, there is a considerable time delay before the reagent is sucked down and dispersed according to the turnover time.

$$\tau_{dp} \cong \frac{V_i}{F_i} + \frac{V_b}{F_i + F_a} * K_z \qquad\qquad (2\text{-}2i)$$

where:

F_i	=	volumetric injection flow (m^3/sec)
F_a	=	volumetric agitator pumping rate (m^3/sec)
K_z	=	geometric factor (1.0 for baffled vertical well mixed tank)
τ_{dp}	=	process dead time from injection and mixing (sec)
V_i	=	injection (piping and dip tube) volume for concentration change (m^3)
V_b	=	broth volume for concentration change (m^3)

The largest source of pH loop dead time is mixing and injection time delays.

The process dead time from mixing is an order of magnitude smaller for DO than pH control. The degree and rate of dispersion for the DO response is the result of both the bubble velocity and the pattern and mass transfer coefficient as set by the agitation speed, sparger design, and injection flow and broth viscosity. The dead time of a clean DO sensor is about four times larger than the dead time for a clean pH sensor, but the net result is that the loop dead time is still much less for DO than for pH.

As a general rule, the loop dead time for DO control is considerably less than the dead time for pH and substrate control.

For a change to be readily distinguished from noise, the excursion in the process variable must exceed the resolution limit and noise band of the measurement. This dead time is on average about one half of the resolution limit or noise band, whichever is significantly larger, divided by the rate of change of the process variable. The resolution limit for an analog-to-digital (A/D) converter with one signed bit is about 0.05 percent and 0.003 percent of span, for a 12-bit and 16-bit A/D, respectively. If the process variable is changing at 0.0001 percent per second (which is not uncommon for the composition response of the liquid phase), then the dead time can be estimated as 250 seconds and 15 seconds for a 12-bit and 16-bit A/D, respectively. The actual dead time can be almost zero, or twice as large as these numbers, because it depends on whether the measurement's starting point is at the opposite or near end of the resolution limit before the change. The dead time from the DO resolution limit is negligible because the DO's rate of change is so fast.

The dead time from the resolution limit of the measurement is variable and significant for slowly changing process variables and 12-bit A/D converters.

In practice, the total loop dead time should be estimated based on the time between an appreciable change in the controller output or set point and the first noticeable change in the measurement. If there are no set point changes with the controller in automatic or output changes with the controller in manual, then you can roughly estimate the minimum dead time as the time delay from the start of the flow set by the batch sequence to the start of the measurement's response.

The time between the start of flow and the start of the measurement's response provides an estimate of the minimum loop dead time.

The observed dead time increases as the rate of change in controller output decreases. This is principally because of the increase in dead time from a slower ramp rate through the final element resolution limit and through the measurement resolution limit and noise band. The sensor may also respond more slowly because of the smaller concentration gradient at the tip of the electrode, particularly if there is any coating. A small set point change has the same effect since it corresponds to a small change in the controller output. The observed dead time for the gradual changes in load caused by the changes in the slope of the growth or product formation rates after the lag phase is much larger than the dead time observed for set point changes. This is because the change in controller output over the loop's response time is miniscule. The control error over the loop's response time, following from equation 2-2d, is incredibly small. The result is even significant fractional improvement in performances has a minimal effect, as equation 2-2b suggests. This explains why even with far less than optimal tuning, DO and pH loops still draw a straight line. The principal advantage gained from improved tuning of primary controllers reveals itself as these controllers work to catch up with the increase in demand associated with the five to ten doubling of cells during the exponential growth phase. The substrate concentration rapidly drops, and the substrate feeding is started. Before the exponential growth phase is well established, there is very little oxygen, reagent, and substrate demand, so the primary controllers do not see or have to deal with significant load changes, and secondary set point changes are minimal [1]. Dissolved oxygen controllers tend to cycle at the beginning of the batch because meters and final elements are sized and controllers are tuned for peak oxygen demand.

The improvement gained from improved tuning of primary controllers is most observable after the exponential growth phase is well established.

The steepness of the slope rather than the change in slope of the biomass profile during the exponential growth phase determines how much work the PID controller has to do to meet DO and substrate demand. In other words, the magnitude of the controller output's rate of change (i.e., the size of shift in controller output within a batch phase) is a better indicator of load than a change in the controller's output rate of change. The tuning settings may change at different points in the batch, but the job for the PID is greatest at the controller output's maximum rate of change.

Sluggish tuning of the secondary loops for DO control can have significant consequences at any time in the batch process. This is because it can cause slow and nearly sustained cycling of the DO controller, which is a result of interaction with the flow or speed loops.

Limit Cycles

A resolution limit anywhere in the control loop causes a limit cycle. The period (T_o) of the limit cycle depends on the loop gains and the controller tuning settings, according to equation 2-1j, developed by Hägglund [8]. The limit cycle may be hidden by data compression, attenuated by signal filtering, and disguised by noise:

$$T_o = 4 * T_i * [1 / (K_{mv} * K_{pv} * K_{cv} * K_c) - 1] \qquad (2\text{-}2j)$$

The amplitude of the limit cycle depends on the final element and process gains, the limit cycle period, and the mixing time constant (τ_m), following from equation 2-2k [9].

$$A_o = (S_r * K_{mv} * K_{pv}) * [T_o / (2 * \pi * \tau_m)] \qquad (2\text{-}2k)$$

where:

A_o	=	limit cycle amplitude (%)
S_r	=	signal resolution of final element (%)
K_c	=	controller gain (dimensionless)
K_{cv}	=	controlled variable gain per measurement span (100%/Span PV e.u.)
K_{mv}	=	manipulated variable gain (Flow e.u./%)
K_{pv}	=	process variable gain (PV e.u./Flow e.u.)
T_i	=	controller integral time setting (seconds/repeat)
T_o	=	limit cycle period (seconds)
τ_m	=	mixing time constant for back mixed volume (seconds)

Sustained cycling is not necessarily a bad thing in that it might actually increase mass transfer rates and stimulate biological processes. However, if the peak in the plot of the biomass growth rate or product formation rate versus the process variable is relatively sharp, the variability can result in suboptimal batch profiles.

2-3.   Self-Regulating Processes

The process response is fast and self-regulating for the secondary loops in the bioreactor cascade control systems. Flow, speed, and coolant temperature reach relatively quickly a new steady state in response to a manual change in the secondary controller's output.

> *Secondary loops generally have a fast self-regulating process response.*

Figure 2-3a shows the open loop self-regulating response of the process to a manual change in the controller output. It is termed open loop because the controller is in manual, and it is termed self-regulating because the process decelerates to a new steady state. The process response in figure 2-3a can be described by a steady-state process gain, a loop dead time, and a single time constant. This approximation is called a "first-order" model.

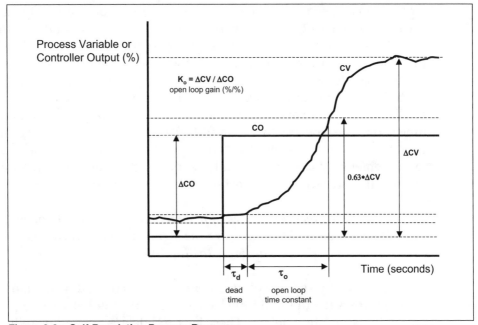

Figure 2-3a. Self-Regulating Process Response

The loop dead time (τ_d) is the time from the start of the change in the controller output to the time at which an excursion of the process variable goes beyond its noise band. The time from the start of the change of the process variable to about 63 percent of its final value is the open loop time constant (τ_o). The final change in process variable (in percent) divided by the change in manual controller output (in percent) is the steady-state process gain (K_p).

Loop dead time delays the controller's ability to recognize and deal with change. If there was no noise, perfect control would be possible if the dead time were zero, in that the controller could immediately see and compensate for a change. The controller would be stable for even the most aggressive tuning settings.

> *If the loop dead time were zero, perfect control would be*
> *theoretically possible.*

For the control of liquid flow and agitation by manipulating a variable speed drive (VSD) with a negligible dead band, the process dead time is small, and most of the loop dead time is set by the module execution time and resolution of the VSD. However, the improvement in control of the secondary loop from PI module execution times of less than one second is overshadowed by other considerations, such as proper tuning of the PI controller. The module execution times of primary controllers should be larger than one second to maximize the signal-to-noise ratio. This is because the real DO, pH, or substrate change over one second is within the noise band. Simply stated, the true change of the process variable within the module execution time should be five times larger than the noise. Larger module execution times are also desirable because they reduce controller loading, but the execution time must be less than twice the oscillation period to avoid severe aliasing. The primary module execution time should also be less than one-fifth the process dead time and time constant, whichever is smallest. The input and output cards and measurement devices are normally set up for over sampling in order to provide five scans for the fastest module execution time.

> *The module execution times of primary loops should be longer*
> *than one second to maximize the signal-to-noise ratio and*
> *minimize the loading.*

A potentially large source of loop dead time for controlling coolant temperature is the thermal time lags for heat transfer between media and piping transportation delays. These lags and delays depend heavily on the coolant system design. Another significant source of dead time is the thermowell lag time, which varies from six to sixty seconds depending on the fluid velocity, fit of the sensing element within the well, and degree of coating [3]. For slowly changing process variables, the dead time that results from the final element's resolution is significant. However, it can be minimized by using sliding stem valves with digital positioners or a special high-resolution VSD [4] [5]. The resolution of a standard variable frequency drive is about 0.1 Hz in 60 Hz, which is about 0.17 percent. The resolution of rotary valves varies from about 0.25 to 5.0 percent depending on the degree of shaft windup and the type of positioner, packing, seating surfaces, and feedback mechanism. The resolution of sliding stem valves that have low friction packing and digital positioners typically ranges from 0.1 percent to 0.5 percent [4].

The heat transfer and transportation delays set during the design of coolant systems are the largest sources of dead time for coolant temperature control.

The open loop time constant can be approximated as the largest time constant in the control loop wherever it is located. The open loop time constant slows down the process variable's excursion rate and gives the controller a chance to catch up with a load change. If a large time constant in the process downstream of the point where disturbances enter is larger than the loop dead time it will improve control by slowing down these disturbances and by effectively filtering out process noise. Stated another way, a small dead-time-to-time-constant ratio improves the potential for good control within the loop. What is achieved in practice depends on the tuning settings, as seen in equation 2-2b. Often, a very large time constant is problematic because open loop tests that wait until a steady state is reached take too long. The time to reach 98 percent of the final response is approximately four time constants plus the dead time, as shown in equation 2-3a. Often the reason a loop is difficult is that the loop is too slow even if the slowness is in the process time constant, which offers the potential for better control.

$$T_{98} = 4 * \tau_o + \tau_d \qquad\qquad (2\text{-}3a)$$

where:
T_{98} =   time to 98% of final response (sec)
τ_d =   total loop dead time (sec)
τ_o =   open loop time constant (sec)

Although a large process time constant might offer improved control of a properly tuned single loop controller, it can cause problems if it is in the secondary loop of a cascade control system. If the time constant in a secondary loop is not significantly less than the time constant in the primary loop, a fundamental rule is potentially violated: the secondary loop must be much faster than the primary loop. Some negative consequences can be avoided by using tuning methods that make the secondary loop faster. The section on cascade control in chapter 3 addresses this concern in detail and provides a solution.

Although a large process time constant can improve the potential for better loop performance, it creates tuning and cascade control issues.

A large time constant caused by a sensor lag time, damping adjustment in the transmitter, or filter on the process variable in the control system also smooths out noise. However, it attenuates the control system's view of the true changes in the process. A large time constant in the measurement path creates an illusion of better control [3] [4] [5]. It also delays the control loop's ability to recognize a load upset.

Measurement time constants (e.g., filter times) give an attenuated version of the real process variability and an illusion of better control.

A controller that has too high a gain or rate action inflicts disturbances on itself by increasing the noise in the controller output. You can determine whether noise in the controller output is insufficiently smoothed out by the process time constant by determining whether the noise with the controller in automatic is greater than the noise when the controller is in manual. The solution to reduce noise introduced by automatic action is to first eliminate excessive gain or rate action and then if necessary increase signal filtering [5].

Noise introduced by automatic action should be first reduced by better tuning and then if necessary by adding a minimal filter on the process variable.

In reality, there is more than one time constant, which gives rise to a bend in the initial response, as seen in figure 2-3a. In the bend, the process variable accelerates. Subsequently, the response reaches an inflection point at which point it decelerates. If there was a single time constant, the response immediately after the dead time would be fastest and would form a right angle. For a "first-order" approximation, the small time constants effectively become dead time. This is determined by the intersection of the process variable's initial value by a line that is tangent to the inflection point, as shown in figure 2-3a.

In real applications, there are many small time constants that can be approximated as additional dead time.

Adding rate action can compensate for these small time constants if the measurement noise is small and a large process time constant exists to smooth out the reaction to noise in the controller output. Flow and speed loops have too much noise and no appreciable process time constant. Composition loops may have too much sensor noise. The best candidates for rate action are generally temperature loops that have A/D noise. When one-way cooling is used (no heating via a tempered water system), then it is essential that derivative action be used to prevent overshoot of the set point.

> *The best candidates for rate action are generally temperature loops with minimal noise, in which the rate time is set to compensate for small time constants.*

The process gain is more linear for flow and speed loops that have a VSD since these loops do not have the flow characteristic of a control valve. The process gain for gear pumps is relatively constant and does not vary much with operating point or process conditions. The steady-state process gain for a temperature loop that throttles coolant flow is inversely proportional to coolant flow. The equal percentage flow characteristic of a control valve can largely cancel out this process gain nonlinearity. If a VSD, linear valve, or secondary flow loop is used, then there is no inherent compensation. At low loads, the process gain increases dramatically. Also, at low flow, the transportation delay and heat transfer lag increase. This combination of a higher process gain and larger loop dead time can cause a limit cycle in a temperature control system. Whether it is noticeable or not depends on data compression, mixing, and tuning.

If the flow loop manipulates a VSD without velocity limits instead of a control valve as the final element, then the time constant is dramatically faster. The loop's speed of response is mostly set by the module execution time, transmitter damping, and signal filters.

2-4.   Integrating Processes

The process response is slow and integrating for the bioreactor primary loops for DO, pH, and substrate control. The concentration control of batch operations that have slow reaction rates resembles an integrating response because no discharge flow occurs for continuous control. Also, in batch operations, there is no steady state as evidenced by the batch profiles. After the dead time, the process slowly ramps in response to a manual change in the primary controller output, as shown in figure 2-4a. There is no steady state and consequently no steady-state process gain for these processes. The process gain is an integrating gain that is the change in ramp rate in %/sec per percent change in the controller output. The response is modeled as a loop dead time and an integrating process gain. Note also that the initial response is not flat for manual operation as it was for a self-regulating process. This is because a small unbalance (difference between load and controller output) causes the process variable to ramp. Thus, the process gain must be calculated from the change in ramp rate.

Primary loops typically have a slow integrating type of process response.

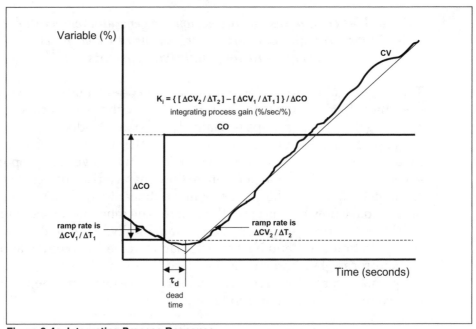

Figure 2-4a. Integrating Process Response

An integrating process does not line out when the controller is in manual.

If the process output's ramp rate (slope) is used for the controlled variable where the process gain is the change in the process output's ramp rate (slope) for a change in the process input, then the result can be modeled as a self-regulating response. When the ramp rate is constant for a given change in the controller output, there is effectively a steady state. However, this ramp rate changes as the batch progresses.

An integrating process can be modeled as a self-regulating process with an interim steady state by computing the change of the process output's ramp rate in response to a change in a process input.

Integrating processes can be identified much faster because there is no steady state. The person or software must just wait until the ramp rate is recognizable, which is usually possible after four dead times.

> *Integrating processes can be identified faster than self-regulating processes.*

Disturbances also change the ramp rate. The dead time and process gain change in relation to operating point and process conditions. The dead time may also change as a function of the direction of the change. For these and other reasons, tests should be repeated at least twice in both directions at different conditions.

> *Tests to identify process dynamics should be repeated at least twice in both directions at different operating points, process conditions, and batch times.*

Though the bioreactor temperature loop may have a self-regulating response, the time required to reach steady state is so long and the self-regulation so weak because of a lack of continuous discharge flow that it is best to model and tune the response as an integrating process. This can be visualized by noting that the first portion of the response in figure 2-3a after the dead time is similar to the response in figure 2-4a. Processes that do not have a true integrating response but ramp in the time frame of interest are called "pseudo" or "near" integrating processes [3] [5]. Equation 2-4a shows the conversion of the steady-state open loop gain (often just referred to as process gain) and the open loop time constant of a self-regulating process into the equivalent integrating process gain of a pseudo integrating process [3] [5]. Modeling a slow self-regulating loop as a pseudo integrating process can save considerable test time. Chapter 3, on basic feedback control, discusses how to apply the tuning rules for integrators to these processes.

> *Batch operations tend to have an integrating type of concentration and temperature response because there is no continuous discharge flow or steady state.*

$$K_i = K_o / \tau_o \qquad\qquad (2\text{-}4a)$$

$$K_o = K_{mv} * K_{pv} * K_{cv} \qquad\qquad (2\text{-}4b)$$

where:
K_i =   "pseudo" or "near" integrating process gain (%/sec/%)
K_o =   steady-state open loop gain (also known as process gain) (%/%)

K_{cv} = controlled variable gain per measurement span (100%/Span PV e.u.)

K_{mv} = manipulated variable gain per pump or valve (Flow e.u./ 100%)

K_{pv} = process variable gain (PV e.u./Flow e.u.)

τ_o = open loop time constant (also known as process time constant) (sec)

The bioreactor pressure loop has an integrating type of response if the loop manipulates a fan VSD or if the pressure drop across the control valve is so large that a change in vessel pressure does not translate into an appreciable change in vent flow. In other words, a higher vessel head space pressure does not force out more off-gas flow through the vent system.

As was the case for the self-regulating response, the initial bend in the response is associated with small time constants. If the process could be modeled by a loop dead time and an integrating process gain, then the process variable would abruptly reverse immediately after the dead time. In real applications, the transition is smoothed by small time constants. A rate time can be set to compensate for these time constants if the measurement noise is not excessive and the integrating process gain is low, which, according to equation 2-4a, is equivalent to a large process time constant.

References

2-1.   Alford, Joseph S., Email correspondence, 2006.

2-2.   McMillan, Gregory K., *Biochemical Measurement and Control.* Reprint via ProQuest UMI "Books on Demand", 1987.

2-3.   McMillan, Gregory K., *Tuning and Control Loop Performance.* 3d ed. ISA, 1992.

2-4.   Blevins, Terrence L., McMillan, Gregory K., Wojsznis, Willy K., and Brown, Michael W., *Advanced Control Unleashed: Plant Performance Management for Optimum Benefits.* ISA, 2003.

2-5.   McMillan, Gregory K., *Good Tuning: A Pocket Guide.* 2d ed. ISA, 2005.

2-6.   McMillan, Gregory K., and Cameron, Robert, *Advanced pH Measurement and Control.* 3d ed. ISA, 2005.

2-7.   McMillan, Gregory K., Sowell, Mark, and Wojsznis, Peter W., "The Next Generation – Adaptive Control Takes a Leap Forward." *Chemical Processing*, September 2004.

2-8.   Hägglund, T., "A Control Loop Performance Monitor." *Control Engineering Practice* 3, no. 11 (1995): 1543-51.

2-9.   McMillan, Gregory K., "What Is Your Control Valve Telling You?" *Control Design*, May 2004, 43-48.

Chapter 3:
Basic Feedback Control

Chapter 3

Basic Feedback Control

3-1. Introduction

Adding feedback measurement is essential for important process outputs because of the uncertainties and variability both in the process inputs and within the process. This is particularly true for bioreactors because of the unknowns in the metabolic pathways and kinetics of the cells. Using measurement in a feedback control loop offers automatic compensation. Since the proportional-integral-derivative (PID) controller is the predominant controller used in industry for basic feedback control, this chapter focuses on how to set up, tune, and optimize the PID.

Chapter 2 discussed how a control loop's ultimate performance depends on a process model but that the actual performance is determined by the PID controller's tuning settings. This chapter details the procedure for computing the PID tuning settings from the parameters of a process model. Though a dynamic model is implied in the tuning, as long as the PID controller is stable, it corrects for unknowns, loads, and disturbances.

In general, the user backs off from the "hottest" tuning settings for the "tightest" control in order to reduce the potential for oscillations in the process variable or manipulated variables. A tradeoff must always be made between performance and robustness. High controller gains transfer more variability from the process variable to the manipulated variable. Fortunately, bioreactor loops have fewer interactions than do many other unit operations, so variability in the manipulated variable is less disruptive. However, fluctuations in the controller output as a result of process or measurement noise that exceeds the final element's resolution can disturb the loop. Aggressive control makes the loop more vulnerable to oscillations from the inevitable changes in the process dynamics. Also, operators tend to dislike the large kicks in the controller output that can occur from set point changes to controllers with a high gain even though these kicks may in fact be beneficial in terms of a creating a faster response.

This chapter shows how process dynamics permit controller gains for the bioreactor's primary loops that are higher than those users have used on loops in other unit operations. In practice the controller gains actually used are far below the optimum for tight control because of the user's comfort zone combined with concerns about robustness and amplification of noise.

The oscillations caused by hot controller tuning and noise amplification are relatively fast and better understood than are the slower oscillations from less obvious causes. This chapter details how sluggish tuning of secondary controllers causes cycling in a cascade control system, how low controller gains cause reset cycling, and how sluggish tuning settings aggravate the observed limit cycling from the final element's resolution limit.

PID controllers are commonly used in bioreactors for dissolved oxygen, pH, and temperature control. Before tuning tests are done and tuning decisions made, users need to select the proper structure and form of the PID algorithm. The effect that any set of tuning settings will have on loop performance depends on the PID's structure and form. A modern distributed control system (DCS) offers two or more forms and eight or more structures. It's important for users to understand the functional capability of each choice in order to standardize on the best algorithm for the application. When migrating projects or relocating settings from laboratories or pilot plants to production units, it is important to know the effects of different controller algorithms and the tuning parameter units. This knowledge enables users to reuse what they learned from tuning previous systems or similar applications in current systems.

Loop tuning and analysis tools also need to take into account tuning setting units, PID structure and form, and the limitations on the testing of production systems. The test times and step sizes are often beyond what is acceptable to plant operations. This chapter discusses how test time can be reduced by an order of magnitude for the primary loops, considerations for using the tuning settings from bench-top units and pilot plants, how to evaluate batch set point responses, and tuning methods that can be extended to provide optimal estimates of the tuning settings for industrial production. Ultimately, the user needs to understand the functional contribution of each mode in order to verify, analyze, and improve controller tuning and performance.

Section 3-2 describes the relative importance of various PID forms and structures; issues to consider when moving tuning settings between bench tops, pilot plants, and production units; and the important advantages offered by external feedback. Section 3-3 develops a unified approach for tuning controllers, discusses how tuning methods can be reduced to a common form, details the impact of process dynamics on primary controller tuning, shows how to greatly reduce process testing time, and reveals how the cell's response prevents direct feedback control by either PID or MPC. The section concludes by introducing an innovative translation of the controlled variable that inherently overcomes these fundamental problems and makes possible batch profile control of biomass and product concentrations.

Section 3-4 provides an overview of an adaptive controller that identifies the process model from normal set point or output changes and eliminates the need for manual or programmed test sequences. The adaptive controller computes settings based on the users' preference of robustness versus performance and users' concerns over the amplification of noise. The tuning settings identified from past batches can be used to schedule (predict) settings as a function of controller output or any process or computed variable such as batch time. Limits can be placed as necessary on the range of settings. Whether these settings are used or not, the models identified provide process knowledge and diagnostics. Changes in the process model can alert operations staff to changes in bioreactor behavior.

Section 3-5 compares conventional techniques for pre-positioning controller outputs and tuning controllers against a new innovative approach that considerably increases the speed of the primary loop's set point response through a simple calculation of the controller output's optimal switching time and final resting value.

Learning Objectives

A. Be able to select the best PID structure and form for an application.

B. Know how to tune PID controllers using a unified approach.

C. Recognize how diverse tuning methods can be reduced to a common form for tight control.

D. Discover how to reduce the process test time by an order of magnitude.

E. Recognize how to prevent fast oscillations caused by aggressive tuning and slow cycling and offsets caused by sluggish tuning.

F. Understand the implications of an integrating response on PID tuning.

G. Learn why biomass concentration and pH must be transformed for optimization.

H. See how an adaptive control can provide process knowledge in addition to automatically identifying tuning settings.

I. Find out how to significantly speed up the set point response to reduce batch time.

3-2.   PID Modes, Structure, and Form

Basic feedback control is performed by a controller that has proportional, integral, and derivative modes. Except for temperature loops, the derivative mode is usually turned off by setting the rate time (derivative time) setting to zero. Users make a distinction here and call a controller with no derivative action a PI controller, but the open literature often does not do this. In many papers, the performance of PI controllers (labeled as PID controllers) is frequently compared to advanced control algorithms. This section will discuss how heavily performance comparisons depend on PID structure. Equations 2-2a and 2-2b in chapter 2 showed how strongly performance depends on tuning settings. The authors of the technical literature choose structures and methods to prove the value of their new tuning methods or algorithms. In cases where derivative control is useful and noise and interaction are negligible, an aggressively tuned PID controller offers the best rejection of unmeasured disturbances at the input to a process [1] [2]. Often overlooked are the special techniques that can be readily added to the PID controller, such as batch preload and dead-time compensation via external reset (mentioned in this section) and the optimal switching of PID output to its final resting value (detailed in section 3-5).

Proportional Mode Structure and Settings

The discrete contribution that the proportional mode makes to the controller output for the "standard" form of the PID algorithm is shown in equation 3-2a. The set point is multiplied by a β factor that ranges between 0 and 1 and is used to provide a proportional kick to speed up the response to a set point change. The kick is a step change in output. For slow loops, the step drives the output beyond its final resting value. Without this kick, the PID controller output relies on the integral action, which provides a slow approach to the set point by way of a slow reset time that is set to match the slow process response. The benefit the kick provides is greatest for loops in which the process time constant or integrating process gain is much slower than the process dead time. Thus, the kick provides minimal benefit to a flow loop since the process time constant is so fast. However, if the flow loop has a cascade set point or remote cascade set point, the kick can be used to provide a more immediate response to the demands of a primary controller or a batch sequence, as discussed in sections 3-3 and 3-5, for cascade and batch control, respectively. However, whenever the operator changes a set point, this kicks the output. For a primary loop or a single loop, a set point rate limit or filter can be added to smooth out the kick from an operator set point or the β factor can be set to 0. It's not advisable to use set point rate limits and filters for secondary loops because they degrade the primary loop's performance. They have the same effect on the controller output as a velocity limit or filter. Although the β factor in a secondary loop degrades a primary loop's ability to reject disturbances, it has no deteriorating effect for disturbances that originate within a secondary or single loop.

A structure of "PI action on error," which sets the β factor = 1.0, provides the fastest set point response. This is important for secondary loops and the batch sequences.

In the equations for the PID controller used in this book, we use the term *controlled variable* (CV) in place of *process variable* (PV) both to denote that the units are percent of scale range instead of engineering units of the process variable and to make the nomenclature for the PID and model predictive controller (MPC) more consistent. It is important to remember that the configuration, displays, trend charts, and documentation of most modern control systems use process variable (PV), feedforward variables, and controller outputs in engineering units. When computing the process gain, users must take this into account because in a distributed control system (DCS) the PID algorithm is based on feedback and feedforward inputs and an output in percent of the respective scales. Special-purpose and user-constructed PIDs, as well as a few programmable logic controllers (PLC), use engineering units in the algorithm. This severely reduces the portability of tuning settings and the users' understanding of the effects of the process. For example, if you use percent in the algorithm, you can tune a flow loop with a gain of 0.2 for most applications that have decent valve sizing and performance. If you used engineering flow units, a loop with a 1-liter-per-minute span would have a controller gain setting that is sixty times smaller than a loop with a 60 kg per hour span. Section 3-3 on PID tuning methods shows that the change in engineering units cancels itself out for PID algorithms that use percent for the controller inputs and outputs. The specification of engineering units for controller output in modern DCS and fieldbus systems does not mean these engineering units are used in the PID algorithm. The conversion of controller output from percent to engineering units is done after the PID algorithm.

The use of percent for inputs and outputs in the PID control algorithm increases the portability and comprehensibility of controller gain settings.

$$P_n = K_c * (\beta * SP_n - CV_n) \tag{3-2a}$$

Though most digital controllers use controller gain, proportional band (PB) was once a prevalent tuning setting for the proportional mode in analog controllers. Proportional band was devised to be the percent change in the controlled variable that is needed to cause a 100 percent change in the controller output [2] [3]. Some digital controllers still use it. It is critical that the user know whether the proportional mode tuning setting is a gain or proportional band because this can have a huge effect as indicated by the inverse relationship shown in equation 3-2b.

> *It is critical to know whether the proportional mode tuning factor is a proportional band in percent (%) or a dimensionless controller gain.*

$$PB = 100\% \, / \, K_c \hspace{4cm} (3\text{-}2b)$$

Integral Mode Structure and Settings

The discrete contribution of the integral mode to the controller output for the "standard" form of the PID algorithm is shown in equation 3-2c. The controller gain setting divided by the integral (reset) time setting is multiplied by the error between the set point and the controlled variable for the current execution (n). This result is multiplied by the integration step size (Δt), which is the module execution time, and added to the integral mode result for the last execution (n-1). The result is an integration of the error factored by the ratio of the controller gain to reset time setting. The reset time is the time required for the integral mode to repeat the contribution from the proportional mode. In the ISA standard algorithm, the reset time setting is in seconds per repeat and is simply called "reset." Often the units are stated as just seconds. Some manufacturers use minutes per repeat, or its inverse (repeats per minute), with no change in the use of the term reset. It is imperative that the user know the units of the reset tuning setting.

> *It is critical to know whether the integral mode tuning factor is a reset time (seconds or minutes per repeat) or its inverse (repeats per second or minute).*

$$I_n = (K_c \, / \, T_i) * (SP_n - CV_n) * \Delta t + I_{n-1} \hspace{2cm} (3\text{-}2c)$$

Derivative Mode Structure and Settings

The discrete contribution of the derivative mode to the controller output for the "standard" form of the PID algorithm is shown in equation 3-2d. The set point is multiplied by a γ factor that ranges between 0 and 1 and can be used like the β factor to provide a proportional kick to speed up the response to a set point change. In this case, the kick is a spike or bump, whereas the kick from the β factor is a step. The effect is short term, and the burden is still on the integral mode to change the output enough to accelerate the process variable. Although this jump in the controller output can help the signal get through the dead band or resolution limit of a control valve, the better solution is to reduce backlash and sticktion by using a more precise throttle valve and digital positioner. The γ factor is normally set to zero to prevent the derivative mode from overreacting to operator-entered set points. A set point filter or velocity limit can be used just as it was to reduce the kick from the β factor for single and primary loops. The γ factor, like the β factor, does not affect a loop's ability to reject a load.

> *The γ and β factor do not affect the load response*
> *of a control loop.*

Since the derivative mode provides a contribution that is proportional to the slope of the change, step changes and noise are disruptive. Most controllers have an inherent filter whose time constant is a fraction α of the derivative (rate) time setting; this ensures that the spike in the output becomes a bump. A typical value for α is $1/8$ to $1/10$. Note in the equation 3-2d that for α and γ factors of zero, the numerator can be simplified to just the change in controlled variable multiplied by the controller gain and rate time. The denominator can be simplified to just the execution time. For this case, it is easier to see that the derivative mode provides a contribution that is proportional to the rate of change of the controlled variable ($[CV_n - CV_{n-1}] / \Delta t$].

> *It is critical to know whether the time units of the derivative*
> *mode tuning factor (rate time) is seconds or minutes.*

$$D_n = \frac{(K_c * T_d) * [\gamma * (SP_n - SP_{n-1}) - (CV_n - CV_{n-1})] + \alpha * T_d * D_{n-1}}{\alpha * T_d + \Delta t} \qquad (3\text{-}2d)$$

Summary of Structures
The β and γ factors are sometimes called *set point weighting factors* and are usually found under the category or term structure in the *controller configuration*. A controller in which both factors are adjustable is called a two-degree freedom controller. Other structures have the β and γ factors set equal to 0 or 1. The user can also omit a mode entirely to get P-only, I-only, ID, and PD control with various assigned factors. PI control is achieved by simply setting the derivative (rate) time to zero. In general, the user must not set the controller gain equal to zero in an attempt to get I-only or ID control. Nor should the user set the integral (reset) time to zero in an attempt to get P-only or PD control. Note that using P-only or PD control requires that additional choices be made about how to set the bias and its ramp time. Table 3-3a lists eight choices offered by one major DCS supplier.

Table 3-2a. List of Major PID Structure Choices

1. PID action on error ($\beta = 1$ and $\gamma = 1$)

2. PI action on error, D action on PV ($\beta = 1$ and $\gamma = 0$)

3. I action on error, PD action on PV ($\beta = 0$ and $\gamma = 0$)

4. PD action on error ($\beta = 1$ and $\gamma = 1$)

5. P action on error, D action on PV ($\beta = 1$ and $\gamma = 0$)

6. ID action on error ($\gamma = 1$)

7. I action on error, D action on PV ($\gamma = 0$)

8. Two degrees of freedom controller (β and γ adjustable 0 to 1)

Normally a structure of "D action on PV" ($\gamma = 0$) is used to prevent spikes or bumps from set point changes but "D action on error" ($\gamma > 0$) can be a temporary fix to minimize the effect of valve backlash and sticktion for set point changes.

Algorithms
In the "standard" form of the PID used in most DCS applications, the contributions of the three modes are added to the initial value of the controller output (CO_i) set to provide a bumpless transfer to the execution of the PID algorithm. This initialization of the controller output occurs for transitions from modes in which the output is manually set, the output is tracking an external variable for cascade or override control, or the output is remotely set by batch sequence. Execution of the PID algorithm occurs for the AUTO (automatic), CAS (cascade), and RCAS (remote cascade) modes [2].

Equation 3-2a through 3-2e are a simplified representation of the "positional" algorithm that is predominantly used in industrial control systems [3]. An "incremental" or "velocity" algorithm used by special-purpose supervisory computers developed in the 1970s would compute the change in controller output from each mode for each execution, and add it to the full controller output from the last execution of the PID algorithm. The "incremental" algorithm inherently eliminates bumps, windup, and synchronization considerations. For supervisory computers, the last output was read back as the current set point so the recovery from a failure of execution or the communication link was smooth. However, these and other concerns are now addressed by the use of the back-calculate (BKCAL) feature and initialization calculations of Fieldbus functional blocks. Each control system supplier also has methods for preventing reset windup and coming off of the controller output limits for the "positional" algorithm. The "positional" offers the advantage of using external feedback for improved override control through more effective transitions between the selections of controller outputs. The "positional" algorithm also offers proportional-only (P-only) and proportional-plus-derivative (PD) control and a fixed bias. The "incremental" algorithm has inherent integral action and no fixed bias. Since the incremental algorithm is not recommended, this book will focus on the equations and function of the positional algorithm.

> *The "standard" form of a PID positional algorithm is the most common type used in a DCS and offers P-only and PD control with an adjustable bias.*

$$CO_n = P_n + I_n + D_n + CO_i \tag{3-2e}$$

where:

CO_i	=	controller output at transition to AUTO, CAS, or RCAS modes (%)
CO_n	=	controller output at execution n (%)
CV_n	=	controlled variable at execution n (%)
D_n	=	contribution from derivative mode for execution n (%)
I_n	=	contribution from integral mode for execution n (%)
K_c	=	controller gain (dimensionless)
P_n	=	contribution from proportional mode for execution n (%)
PB	=	controller proportional band (%)
SP_n	=	set point at execution n (%)
T_d	=	derivative (rate) time setting (seconds)
T_i	=	integral (reset) time setting (seconds)
α	=	rate time factor to set derivative filter time constant (1/8 to 1/10)
β	=	set point weight for proportional mode (0 to 1)
γ	=	set point weight for derivative mode (0 to 1)

Figure 3-2a shows the combined response of the PID controller modes to a step change in the set point (ΔSP). The proportional mode provides a step change in the controller output (ΔCO_1). If there is no further change in the SP or the CV does not respond, no additional change in the output occurs even though there is a persistent error (offset). The size of the offset is inversely proportional to the controller gain. Integral action ramps the output unless the error is zero. Since the error is rarely exactly zero, the reset is always driving the output. Even if there are no disturbances, reset action causes a continuous equal amplitude oscillation (limit cycle) as it moves the PID output and the process variable through the resolution limits of the final element and measurement, respectively. The probability of a controlled variable coming to rest exactly at set point is next to zero because of the quantization of the process input by the final element resolution and the process output by the measurement resolution. Thus, a PV nearly always passes through the set point and never stays at a resting value for a controller that has integral action [2].

> *The integral mode continually works to eliminate an offset and is always moving the PID output since the error is rarely exactly zero.*

The contribution made by the integral mode equals the contribution made by the proportional mode ($\Delta CO_1 = \Delta CO_2$) after the integral time. Hence, the integral (reset) time is the time (seconds) it takes to repeat the contribution of the proportional mode, which is the basis of the units "seconds per repeat," for a step change in the SP or CV [2] [3].

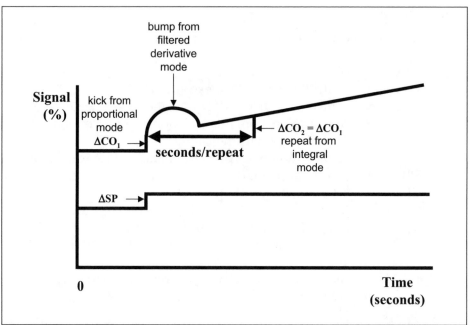

Figure 3-2a. Contribution of Each PID Mode for a Step Change in the Set Point (β=1 and γ=1)

The contribution made by the derivative mode for the same step change is a bump rather than a spike because of the built-in derivative filter. If there is no further change in the SP and the CV does not respond, the contribution from the derivative mode goes to zero [2] [3]. Like the proportional mode, the derivative mode does not attempt to eliminate an offset. Consequently, proportional-plus-derivative (PD) controllers require that a bias be properly adjusted to minimize the offset, particularly for low controller gains.

Proportional-plus-derivative (PD) controllers with low controller gain settings require that attention be paid to the proper setting of the bias to ensure that the offset is acceptable.

Before the advent of the DCS, most industrial control systems used the "series" PID algorithm in which the derivative mode was computed first as the input to the proportional and derivative modes, as shown in figure 3-2b. The computation of the derivative mode in series with other modes was the most practical method for implementing derivative action in analog controllers and was known as the "real" algorithm. In analog controllers, the derivative mode with its built-in filter was actually a lead-lag in which the lead time was the derivative or rate time setting (T_d) and the lag time was the filter time (α^*T_d). In these analog controllers, the derivative action was on PV instead of error ($\gamma = 0$).

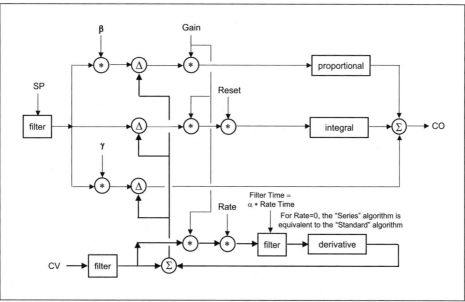

Figure 3-2b. Block Diagram of "Series," "Real," or "Interacting" PID Algorithm

The "series" algorithm was also known as the "interacting" algorithm because the derivative and integral time settings had an interacting effect on the contribution of all modes, as defined in equations 3-2f through 3-2i. From equation 3-2i, the interaction factor would ensure that the ratio of the derivative time to integral time never exceeded ¼. In theory, a ratio of ½ is optimal, but the prevalence of noise makes this goal impractical. A derivative time effectively greater than the integral time setting causes instability. If the user made a mistake and set the derivative time greater than the integral time, the "series" or "interacting" algorithm prevented the effective ratio from exceeding ¼ [3].

The "standard" algorithm computes the derivative mode in parallel with the other modes, as shown in Figure 3-2c. It is the form of the PID algorithm adopted by ISA as its standard. The "standard" or ISA algorithm is also known as the "ideal" or "noninteracting" algorithm and is the default choice in most twenty-first-century control systems. The settings on the left side of equations 3-2f to 3-2h are the settings for the standard algorithm. Note that when the derivative time is zero, the "standard" and "series" algorithm are the same, the interaction factor is 1.0, and no conversion of settings via these equations is needed [3]. Most of the literature on controller tuning assumes a "standard" structure. Hence, the derivative time is often listed as ¼ the integral time for PID control. In section 3-3 on controller tuning, we discuss how derivative action is mostly used on temperature loops; it should be set equal to the next largest time constant (second lag). For thermal systems, the second lag is about 1/10 of the largest time constant.

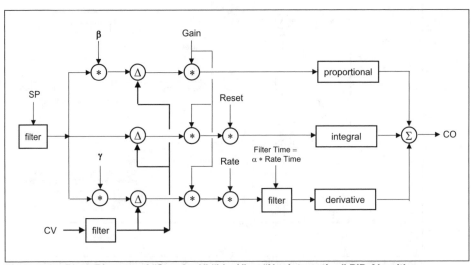

Figure 3-2c. Block Diagram of "Standard," "Ideal," or "Noninteracting" PID Algorithm

Much less common is the "parallel" algorithm, which also computes the modes in parallel. However, the controller gain setting only affects the contribution of the proportional mode. The controller gain is not in the input path to the derivative and integral modes. Only a few systems offer this as a choice. Although there is no performance incentive, some users may prefer that the tuning settings be completely independent. Other users are unappreciative of the drastically different "feel" and settings encountered when tuning a "parallel" controller. Also, the portability between the "parallel" and "standard" systems is much less than the portability between the "series" and "standard" systems. Unless the controller gain is close to 1.0, it is critical that equations 3-2j through 3-2l be used to convert the settings between "parallel" into a "standard" controller when moving settings, regardless of whether derivative action is used.

As always, it is essential to take into account the different units of the tuning settings.

$$K_c = K_c' / I_f \tag{3-2f}$$

$$T_i = T_i' / I_f \tag{3-2g}$$

$$T_d = T_d' * I_f \tag{3-2h}$$

$$I_f = T_i' / (T_i' + T_d') \tag{3-2i}$$

$$K_c = K_c'' \tag{3-2j}$$

$$T_i = T_i'' * K_c'' \tag{3-2k}$$

$$T_d = T_d'' / K_c'' \tag{3-2l}$$

where:

K_c	=	controller gain for "standard" algorithm (dimensionless)
K_c'	=	controller gain for "series" algorithm (dimensionless)
K_c''	=	controller gain for "parallel" algorithm (dimensionless)
I_f	=	interaction factor (dimensionless number < 1.0)
T_i	=	integral (reset) time for "standard" algorithm (sec/repeat)
T_i'	=	integral (reset) time for "series" algorithm (sec/repeat)
T_i''	=	integral (reset) time for "parallel" algorithm (sec/repeat)
T_d	=	derivative (rate) time for "standard" algorithm (sec)
T_d'	=	derivative (rate) time for "series" algorithm (sec)
T_d''	=	derivative (rate) time for "parallel" algorithm (sec)

Instead of an integration of the error for the integral mode calculation, some manufacturers have found it advantageous to use a filtered positive feedback, as shown in figure 3-2d [4]. The input to the filter is either the controller output or an external feedback signal. The output of the filter is added to the net of the proportional and derivative modes in conformance to either the "series" or "standard" form. The filter time is the reset time setting. This positive feedback arrangement facilitates three important features [16]. First, the positive feedback makes it possible to use a "dynamic reset" option in which the external feedback is the PV of a secondary loop or control valve position. This option prevents the controller output from trying to go faster than the velocity-limiting effect of the reset time or the process time constant of the secondary loop, or faster than the slewing rate of a control valve. Second, the positive feedback arrangement inherently prevents a "walk-off" of controller outputs for override control. In the "walk-off," the controller outputs gradually move to an output limit from a continual back-and-forth selection of controller outputs when the integral mode uses an integration of error. The solution for these controllers is to add a filter to each external feedback signal with the filter time set equal to the respective controller's reset time. This effectively creates the same configuration offered by the positive feedback method when the controller is in the integral tracking mode. Third, a dead-time block can be added to the positive feedback path with its dead time set equal to the loop dead time. This provides a dead-time compensator as effective as a Lambda-tuned Smith Predictor but without the adjustments of process gain and lag.

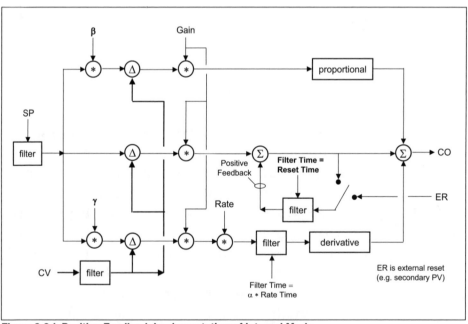

Figure 3-2d. Positive Feedback Implementation of Integral Mode

The positive feedback type of integral mode prevents reset action from outrunning the speed of response of secondary loops and final elements, eliminates walk-off of override controller outputs, and facilitates simplified dead-time compensation.

3-3.   PID Tuning

Focus
The wide spectrum of methods and results for tuning controllers can be bewildering. Most of the published literature analyzes methods for a particular range and type of process dynamics, disturbances at the process's output rather than its input, a set point response, fixed small dead times, and an accuracy of tuning settings not obtainable in industry. In general, the applications are high-speed servomechanism-type responses with measurement noise. Process disturbances enter into the broth as changes in the charges, flows, or metabolic processes of the cell. These load upsets are process inputs that enter the broth just as the manipulated variables do, as shown in the block diagram in figure 2-2a of chapter 2. Disturbances and the manipulated variables are process inputs that change the charge, component, and energy balances. The dead times and time constants associated with the primary loops' response are large and variable. Continual load upsets, the moving target of a batch profile, the nonlinearities enhanced by changing batch conditions, variable dynamics, resolution limits, and process noise from imperfect mixing place severe practical limits on the repeatability of tuning settings.

Control theory centers on high-speed servomechanism
response with noise.

This section focuses on the Lambda method, showing how it can be adjusted to meet process goals and giving when desirable an equation similar to other tuning methods that are touted for loop performance. A translation of form also enables a much faster test time. The Lambda method has almost become a universal method because of its fundamental design.

The Lambda tuning method can be adjusted or transformed to
meet any type of process objective and offers a unified
approach to controller tuning.

Most of the articles to date on loop tuning are relevant to self-regulating processes with dead-time-to-time-constant ratios in the range 0.5 to 2.0, set point changes or step disturbances entering into the process output whose effect is similar to the set point changes for the most common type of controller (PI with $\beta=1$), and a presupposed accuracy of tuning settings of 10 percent or better. When integrating processes have been evaluated for tuning, it has typically been a level loop on a surge tank.

*Most of the literature on tuning deals with a narrow range of
dynamics, self-regulating processes, and a set point response
for PI control action on error (β=1).*

In the process industry, 99 percent of the self-regulating control loops
have dead-time-to-time-constant ratios that range from 0.05 to 20.0. A true
or "near integrating" type of response is prevalent in primary loops for
concentration, pressure, and temperature control. The integrating process
gain is usually extremely small and is equivalent to a very large process
time constant. Process load upsets and disturbances occur at the process
input instead of at the process output as commonly shown in the
literature. Consequently, in industrial applications a process variable's
rate of change from an upset or disturbance usually depends on the
process time constant or integrating process gain. For bioreactors, the rate
of change is very slow.

These same loops exhibit a variability of 50 percent or more in tuning
settings from process tests as a result of noise, nonlinearity, resolution
limits, and unmeasured disturbances [2]. A repeatability of 25 percent of
tuning test results is considered exceptional. Realizing tuning setting
precision in industrial applications eliminates a lot of the hype associated
with methods and software. It also prevents false expectations, which is
particularly important since engineers and scientists are accustomed to
computing numbers to two or more digits [5].

*Differentiating tuning methods and software based on the
second digit of the tuning setting computation have little
value in industry.*

Temperature Loops
Temperature loops commonly have a process dead time and second time
constant that are each about 1/10 of the largest time constant because of
the interactive thermal lags [6]. A second-order model (two time
constants) is useful for identifying the derivative time setting. However, in
practice, it is difficult to accurately identify this second time constant so a
first-order model (one time constant) is often used. This is often adequate
for temperature loops since about half of the unidentified second time
constant ends up as additional dead time and half ends up as an
incremental increase in the largest time constant. The dead-time-to-time-
constant ratio is about 0.1 whether a first- or second-order model is used.

*The dead-time-to-time-constant ratio for most temperature
loops on mixed volumes is about 0.1 whether a first-order or
second-order model is used.*

Because processes with large time constants and small dead times appear to ramp in the region of interest, the difference between a self-regulating and integrating process response is blurred and the distinction becomes a matter of convenience. If the process is visualized as integrating, the test does not have to wait until the process variable plateaus to a new final value. This is particularly advantageous for loops that have large time constants, since it takes about four time constants plus the dead time to reach steady state. If the user prefers to use tuning equations for self-regulating processes, the integrating process dynamics can be converted into self-regulating process dynamics, and vice versa.

The lack of a liquid discharge flow in a batch operation reduces the self-regulation of the temperature loop. However, for a bioreactor that has a well-designed coolant, the change in driving force across the heat transfer surface means that the temperature loop is not a pure integrator. For example, if the temperature change is large enough, the temperature difference between the broth and coolant should be large enough to change the heat transfer to the coolant sufficiently for the temperature to reach a steady state. However, the temperature change may be much larger than the desired change in temperature and the time to reach a steady state too slow.

> *Bioreactor temperature loops have a time constant so large that the distinction between a self-regulating and integrating response is a matter of test time requirements and computational preferences.*

Secondary Loops
In a cascade control system the output of the primary controller, such as bioreactor temperature, is the set point of a secondary controller, which helps linearize the loop and compensate for disturbances that affect the final element's ability to do its job. Secondary flow and speed loops have a dead-time-to-time-constant ratio of between 1.0 and 4.0 unless a large signal filter has been added to measurement or there is significant velocity limiting in the drive or valve response. Most of the dead time in these loops comes from the module execution time and the resolution of the drive or valve response. Secondary coolant temperature loops have a much larger process dead time and time constant whose ratio and magnitudes significantly vary with the coolant system design.

Closed and Open Loop Responses
The terms *closed loop* and *open loop* are commonly used in control and are essential to understanding Lambda tuning. The term *closed loop* is used to denote that PID control action is active (PID is in AUTO, CAS, or RCAS modes). This means it is changing a manipulated variable in response to a process variable, effectively closing the signal path between the CV and

CO of the PID in the block diagram of a control loop (see figure 2-2a). Conversely, the term *open loop* is used to denote that PID control action is suspended (PID in IMAN, MAN, or ROUT) and that the signal path is open between the CV and CO. The open loop response is the response of the process only. The closed loop response is the combined response of the process and PID and shows the effect of the control algorithm and tuning settings [2].

Figure 3-3a shows the open loop time constant (τ_1) that is created by making a step change in the controller output when the controller is in manual for a self-regulating process. Figure 3-3b shows the closed loop time constant known as Lambda (λ) that is created by making a step change in the controller set point when the controller is in automatic. In both cases, the time constant is the time it takes the controlled variable to reach 63 percent of its final value after the process variable starts to change (after the loop dead time). In tuning methods in which the time constant is estimated as the time required to reach 98 percent of the final response (T_{98} as defined by equation 2-2a in chapter 2) divided by 4, the resulting time constant includes ¼ of dead time. In actual applications, there is no sharp transition from the negligible response caused by loop dead time and the exponential response associated with the primary time constant. The smoothing that occurs at the beginning of the exponential response is caused by a smaller second time constant (τ_2). Though this small time constant is difficult in practice to identify, it can be estimated for temperature and many concentration loops on well-mixed volumes as being 1/10 of the largest time constant.

The open loop time constant is the largest time constant in the loop. If it is in the process between the point of entry of the disturbances and the sensor, the open loop time constant slows down the excursion caused by a disturbance. This gives the controller a better chance of catching up to the disturbance. If the largest time constant is in the measurement, it provides an attenuated view of process variability and slows down the reaction to unmeasured disturbances. Coated electrodes can cause the sensor time constant to approach or even exceed the process time constant. Also, in the case of a liquid flow or speed loop, a transmitter damping setting or process variable filter time setting that is greater than two seconds generally means the largest time constant is in the measurement. This is because the liquid flow process and final element time constant is less than two seconds [2].

Lambda Tuning
The ratio of the closed-loop-to-open-loop time constant is the Lambda factor, as in equation 3-3a. For secondary flow and speed loops, a Lambda factor of 2 ($\lambda_f = 2$) is a good starting point. This gives a closed loop time

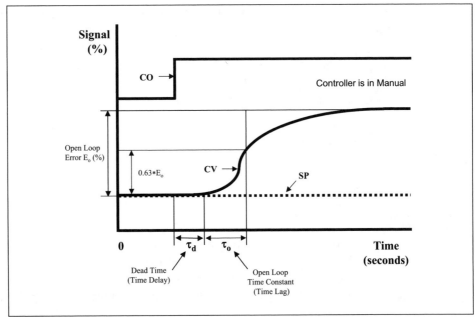

Figure 3-3a. Open Loop Time Constant

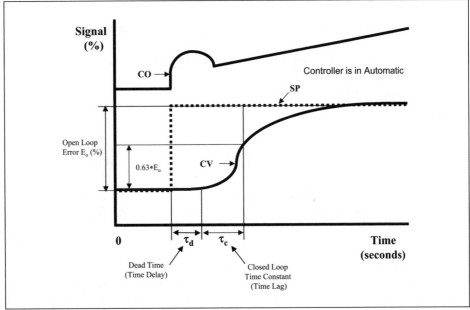

Figure 3-3b. Closed Loop Time Constant

constant that is twice the open loop time constant. For primary loops, the Lambda factor is set less than one to provide tighter loop control.

$$\lambda = \lambda_f * \tau_1 \qquad (3\text{-}3a)$$

$$T_i = \tau_1 \qquad (3\text{-}3b)$$

$$K_c = \frac{T_i}{K_o * (\lambda + \tau_d)} \qquad (3\text{-}3c)$$

$$T_d = \tau_2 \qquad (3\text{-}3d)$$

where:

K_c = controller gain (dimensionless)
K_o = open loop gain (also known as process gain) (%/%)
λ = Lambda (closed loop time constant) (sec)
λ_f = Lambda factor (ratio of closed to open loop time constant) (dimensionless)
τ_d = total loop dead time (sec)
τ_1 = largest open loop time constant (also known as process time constant) (sec)
τ_2 = second largest open loop time constant (sec)
T_i = integral (reset) time setting (sec/repeat)
T_d = derivative (rate) time setting (sec)

For derivative (rate) action to be effective, the loop response should be smooth and have a significant second time constant [3]. This is generally the case for temperature loops on mixed volumes. Since the second time constant is about 1/10 of the largest time constant, the rate time setting for temperature loops is about 1/10 of the reset time setting.

Derivative action that has a rate time setting of about 1/10 of the reset time setting can provide tighter bioreactor temperature control.

Contribution of Loop Components
The open loop gain is the product of the steady-state gains for each major component in the loop, that is, the final element (manipulated variable), process piping and equipment (process variable), and sensor and transmitter (controlled variable), as shown in equation 3-3 e. The result must be dimensionless (%/%), and if it is not, a component is missing or the engineering units are not consistent. The manipulated variable gain is generally more linear for variable speed drives than for control valves. The process gain is usually nonlinear for temperature and concentration loops, especially pH. The process gain is 1.0 for flow. The measurement

gain is linear and provides a conversion from the engineering units of the process variable into percent of the measurement span (K_{cv} =100%/ EUspan) [2] [3] [7].

$$K_o = K_{mv} * K_{pv} * K_{pv} \qquad\qquad (3\text{-}3e)$$

where:

K_o =   open loop gain (%/%)
K_{cv} =   controlled variable (measurement) gain (%/EU)
K_{pv} =   process variable (process) gain (EU/EU)
K_{mv} =   manipulated variable (final element) gain (EU/%)

Note that in the literature on control, no analysis is usually given of the contribution of loop components to the open loop time constant or the open loop gain. In fact, everything outside of the controller is often referred to as the process, and there is no consideration or even appearance on a control diagram of final element and instrumentation details or location. Consequently, the open loop time constant, open loop gain, and the total loop dead time are called the *process time constant*, *process gain*, and *process dead time*, respectively. The total loop dead time is the summation of all pure dead times and time constants smaller than the second-largest time constant, no matter where they appear in the loop.

The open loop gain is commonly called a process gain despite the fact that it is dependent on the product of the final element, process, and measurement gains.

The open loop time constant is commonly called a process time constant even though it is the largest time constant in the final element, process, or measurement.

The total loop dead time is commonly called a process dead time even though it is really the summation of all time constants smaller than the second-largest time constant and the dead times in the final element, processes, and measurement.

Unified Approach

Since in bioreactors there are few interactions and small changes in controller outputs can cause resolution delays, smoothing the manipulated flow may be less important than the ability to control a primary loop at its set point. For the tightest control, the tuning is set to transfer as much variability from the process variable (controlled process output) to the controller output (manipulated process input) as possible. Suppose there is a Lambda factor of 0.1 and a dead-time-to-time-constant ratio of 0.1, as commonly found in temperature loops. In this case, the

Lambda tuning equation reduces to ½ the open loop time constant divided by the product of the open loop gain and total loop dead time. The result is equation 3-3f, which is the simplified internal model control (SIMC) equation for tight control. Most documentation on Lambda tuning has focused on slowing down the response of fast loops so as to provide a smoother and a more consistent response, which is important for improving the coordination or reducing the interaction between loops on unagitated volumes. However, nothing prevents the Lambda tuning factor from being set less than 1.0 in order to provide much tighter control. Furthermore, unlike most other tuning knobs, setting the Lambda factor for more aggressive control does not cause the loop to become unstable. Rather, the loop just approaches the tuning equations cited for best load rejection capability. In fact, most of the equations developed in the 1960s and 1970s that focused on excellent load rejection have this common form of the controller gain, in which the gain is set equal to some factored ratio of the process time constant to the product of the open loop gain and total loop dead time [2]. Appendix C, "Unification of Controller Tuning Relationships," in this book, shows how Ziegler-Nichols, Lambda tuning, and internal model control tuning equations reduce to this common form. This reduction confirms diverse methods and opens the door for a unified approach to tuning [8].

$$K_c = 0.5 * \frac{\tau_o}{K_o * \tau_d} \qquad (3\text{-}3f)$$

For tight control, most of the tuning settings that have been developed are reducible to the common form, in which the controller gain is proportional to the time-constant-to-dead-time ratio.

The Lambda factor should be set based on the size of the open loop time constant and whether the goal is tighter control or better coordination and less interaction between loops. For cascade control, the secondary loop's time response should be five times faster than the primary loop's to prevent interaction between these loops [2] [3]. The ratio of the closed loop time constants of the primary to secondary loop should be greater than five to meet this criterion. If the Lambda factor for the secondary loop is 2, then the ratio should be large enough for even bench-top units. For a flow or speed loop that has a two-second time constant, the secondary loop Lambda would be four seconds, which means the primary loop Lambda should be at least twenty seconds. This is not a problem for temperature, but could be a consideration for a pressure loop that is near the end of a batch when the off-gas flow is high.

Pressure Loops

The pressure loop has some self-regulation from its off-gas flow. However, the change in pressure in the fermenter may not translate into a significant change in pressure drop if the vent system pressure is much lower than the bioreactor pressure. Also, the final element may be a fan rather than a vent valve, which makes the change in vent flow even less dependent on fermenter pressure. Since the pressure change and time required to reach steady state are beyond the frame of interest, the pressure response can for all practical purposes be considered a ramp. The process time constant is the head space and gas holdup volume divided by the off-gas flow rate. The process dead time is less than a second, which means the total loop dead time is largely determined by the final element's resolution limit and the noise band and dynamics of the measurement. For a well-designed pressure loop the dead-time-to-time-constant ratio is less than that for the temperature loop and, like the temperature loop, can be modeled as if it were an integrator. Loops that can be treated as integrators that are really self-regulating if control and time limits are nonexistent are called *near integrators* or *pseudo integrators* [2] [3].

Near Integrators

For "near" integrators, the integrator gain is the steady-state open loop gain divided by the open loop time constant, as shown in equation 3-3g. Note that the units of %/% per second or 1/second are better understood as the %/sec change in the ramp rate of the CV per percent change in the CO for the controller in manual. Since the lack of a steady state means that the process ramps when the controller is in manual, the integrator gain is measured as the change in ramp rates from before to after the change controller output. This method is also known as the "short cut method" and is an extension of the process reaction curve method developed by Ziegler and Nichols [3] [8] [9]. The dead time is the time from the change in controller output to a significant change in ramp rate. Normally, the output change is in the direction that will change the direction of the ramp, which means the dead time is the time until the reversal of slope [3].

$$K_i = \frac{K_o}{\tau_o} \tag{3-3g}$$

$$K_i = \frac{CV_2/\Delta t - CV_1/\Delta t}{\Delta CO} \tag{3-3h}$$

If we substitute equation 3-3g into equation 3-3c, we have the controller gain for a "near integrating" process [2] [3] [8]. The tuning settings can now be determined from the change in ramp rates per % and total loop dead time and a desired closed loop time constant (λ). The tuning method or software does not need to wait for the time to steady state but can usually identify the new ramp rate in five loop dead times. If you consider that the loop dead time is 0.1 or less than the process time constant and that it would take five time constants plus the dead time if a steady state could be reached, then the test time to identity the model and hence the tuning settings is reduced by an order of magnitude. For example, the time required identifying the dynamics of a primary loop with a dead time of one minute and a time constant of ten minutes for a step change in controller output would be fifty minutes as a self-regulating response and five minutes as an integrating response. The actual time that software takes may be twice as long as stated to help screen out the effect of noise and disturbances. The test should be repeated in both directions if possible, and it may require performing tests at different batch times for several batches to see how much the dynamics change with batch conditions. For these reasons, the time saved by modeling processes with large time constants as "near integrators" is significant.

Modeling processes that have large time constants as "near integrators" can reduce by a factor of ten the open loop test time for identifying the process dynamics.

$$K_c = \frac{1}{K_i * (\lambda + \tau_d)} \tag{3-3i}$$

If the controller on a process that has a large time constant is properly tuned with a Lambda factor of much less than one, then the time required for a closed loop test is reduced dramatically for the self-regulating model. This is because the closed loop time constant is a fraction of the open loop time constant.

Tight tuning of primary loops with large process time constants makes it possible to much more quickly identify self-regulating models for a closed loop test.

Loop Cycling
"Near integrating" and "pure integrating" processes develop slow rolling oscillations if the integral (reset) time is too small (i.e., too fast), and they have the unusual characteristic of an increasing propensity to develop these oscillations as the controller gain is decreased. In Appendix C, "Unification of Controller Tuning Relationships," equation 3-3j is developed for a reset time that gives zero overshoot (critically damped response) [8]. If a slight overshoot and oscillations with a decay ratio of 1/20 is permissible, then the numerator can be reduced to 0.7 for faster load rejection. A range of 1 to 4 for the numerator is used in practice.

$$T_i > \frac{4}{K_i * K_c} \qquad\qquad (3\text{-}3j)$$

Integrating processes exhibit the following nonintuitive behavior: integral time is reduced if the controller gain is increased because the minimum integral (reset) time setting is inversely proportional to the controller gain.

Primary loops that have too large a controller gain develop oscillations that have a period of close to the ultimate period (e.g., four dead times). This is much faster than the slow rolling oscillations that are caused by combining a low controller gain and low reset time [2] [3]. Ultimate oscillations are uncommon in bioreactor loops because the maximum gain (ultimate gain) that would trigger these oscillations is quite high for well-designed primary loops. Relatively fast oscillations in the primary loop often indicate a piping or mixing problem. More common if not hidden by historian data compression is the limit cycle from the resolution limits of the final elements and measurements (figure 3-3c), the slowly decaying oscillations caused by a reset setting too fast for a controller gain setting (figure 3-3d), or secondary loops that are tuned for too slow of a response (figure 3-3e).

Most of the oscillations in primary loops result from limit cycles caused by resolution limits, slowly decaying cycles from too much reset action, and interaction with secondary loops.

Most of the "near integrator" processes have a large process time constant and a small process gain. Consequently, integrator gains of about 0.001 to 0.1 and 0.00001 to 0.001 %/sec per % are common for bioreactor pressure and temperature loops, respectively. For the temperature loop, tuning the controller with maximum gain for a dead time of 100 seconds would correspond to a controller gain of greater than 10, which is beyond the comfort zone of most users. Consequently, a more moderate controller gain is used that transfers as much of the variability from the controlled variable to the manipulated variable as desired. Since this controller gain is often significantly less than the maximum allowed by equation 3-3i, it is critical that equation 3-3j be used to prevent slow rolling oscillations.

Since the primary loop controller gain is set significantly below its maximum, it is imperative that the reset time be increased to prevent slowly decaying oscillations.

Composition Loops
The dissolved oxygen (DO) loop has an integrating response because there is no broth discharge flow and the rate at which oxygen leaves the

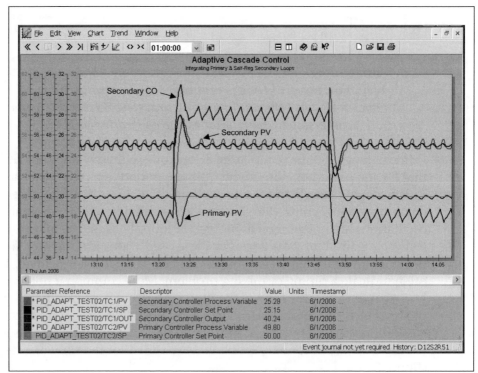

Figure 3-3c. Limit Cycle in Cascade Loop from Final Element Resolution Limit

dissolved phase does not change with dissolved oxygen concentration in the short term. This is because the DO depletion rate is set by the oxygen uptake rate (OUR) of the cells. The OUR of cells gradually changes with biomass growth and product formation rates, but the kinetics are slow. Several ramp rates may be observed for a step change in DO controller output. The first ramp rate is established relatively quickly via the mass transfer of oxygen from the bubbles to the dissolved phase. This first ramp rate is what should be used to tune the controller because it is the time frame of interest. In other words, we want the DO controller to take action and see and respond to changes in the first ramp rate in order to achieve its objective. The second ramp rate via the mass transfer of oxygen from the head space takes longer to show up because the oxygen concentration change in the vapor space is set by the residence time of this volume. Since the mass transfer from the vapor space is normally significantly less than from the bubbles, the effect may not be noticeable except in poorly mixed broths. If the loop is left in manual long enough during the exponential growth phase, then there is a third ramp rate with eventually a negative slope as a result of the increase in oxygen uptake rate with biomass concentration. This decline in DO is the natural batch profile for manual DO control and is often the profile analyzed in the literature for batch fault detection using multivariate statistical process control (MSPC) and principal component analysis (PCA). If the DO controller is in auto and

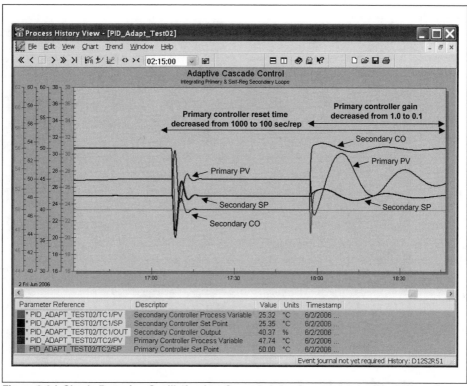

Figure 3-3d. Slowly Decaying Oscillation in a Cascade Loop from an Integrating Primary Loop Reset Time that Is Too Fast per Equation 3-3j

doing its job, the batch profile to be analyzed is the increase in gas feed and agitation rates instead of the DO.

> *The DO open loop response may have several ramps, but the one that is used for tuning is the first one established by the mass transfer from the bubbles to the broth.*

The pH loop has an integrating response because there is no broth discharge flow, and the reagent demand slowly changes along with biomass growth and product formation rate. The S-shaped operating point nonlinearity of the titration curve may make the response look like it is lining out. If the loop is left in manual long enough, the pH eventually drops as the batch progresses because cells normally produce an acid. This decline in pH is the natural batch profile for manual pH control and is often the profile analyzed in the literature for batch fault detection by MSPC and PCA. If a bacterial infection occurs, the drop in pH can be much faster because of the production of lactic acid or acetic acid from the bacteria's consumption of glucose. A pH that is either too low or high causes a decrease in biomass growth rate and reagent demand, which can cause the drop in pH to slow down or even temporarily reverse its direction. Thus, an increase in reagent feed beyond the demand can be

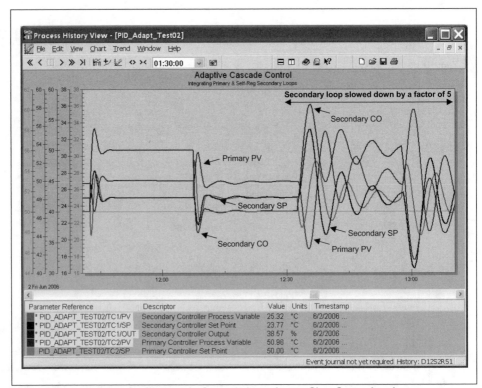

Figure 3-3e. Interacting Oscillations in Cascade Loop from a Slow Secondary Loop

more problematic than a decrease in reagent because it can lower biomass growth. This decreases acid production and accelerates the increases in pH, especially since the slope of the titration curve normally gets steeper as the system approaches 7 pH.

> *The pH loop has a nonlinear ramp rate in which an increase in*
> *a reagent that causes a higher-than-optimum pH when*
> *aggravated by a steeper portion of the titration curve can*
> *cause the positive ramp rate to accelerate.*

The substrate loop, like the pH loop, has an integrating response because there is no broth discharge flow and the substrate demand slowly changes with biomass growth and product formation rate. However, here there is no nonlinearity associated with the operating point, and the substrate demand is much higher than the reagent demand. Consequently, a substrate system that is out of balance, where the difference between the demand and feed rate is significant, can have a faster ramp rate than the pH system. The exception is if a final element with excessive reagent capacity or the acceleration from the steep portion of a titration curve comes into play in the pH loop.

The substrate response normally has a faster ramp rate than the pH response because the potential unbalance (mismatch) between the feed rate and its demand is greater.

It is important to remember that there is a balance point with all integrators, that the degree of imbalance determines the ramp rate, and that the ramp does not change direction until the unbalance changes sign. For temperature, DO, pH, and substrate loops, the coolant, oxygen, reagent, and substrate demand, respectively, determine the balance point. For bioreactor head space pressure, the oxygen uptake rate, carbon dioxide production rate, and off-gas flow set the balance point. The implications of this behavior are explored in section 3-5 on optimizing the set point response.

The ramp rate of an integrating process is proportional to the magnitude of the difference between the feed and the demand or exit flow, and the ramp rate cannot reverse direction until the unbalance reverses sign.

The ramp rate of the integrator response is inversely proportional to the mass holdup, which is the volume for a constant density. Thus, an increase in bubble holdup and head space decreases the ramp rate for pressure control, and an increase in broth volume decreases the ramp rate for the bioreactor temperature and composition loops. Small fermenters have faster ramp rates for the same magnitude of unbalance. Of course, the scaling of flows should match the volume, but the sizing and unbalance may not match. Bench-top units tend to have much faster ramp rates. The change in ramp rate (integrating process gain) must be considered when moving tuning settings to production units.

The ramp rate of an integrating process is inversely proportional to the volume.

Process disturbances for bioreactors are generally very slow because of the volume and the slowness of the kinetics. The loop's performance for slow disturbances depends more on integral action than on proportional action. The rate of change of the controller output needs to effectively exceed the load disturbance's rate of change in order to return the PV to set point. Figure 3-3f shows how a small persistent offset develops in response to a second upset that is the same size as the first upset but ten times slower.

Biomass and Product Profile Control

Biomass and product concentrations may be measured on line but are not often directly controlled. First-principle models enhanced and adapted by neural networks as well as model predictive control offer the ability to make faster and more reliable inferential measurements of these

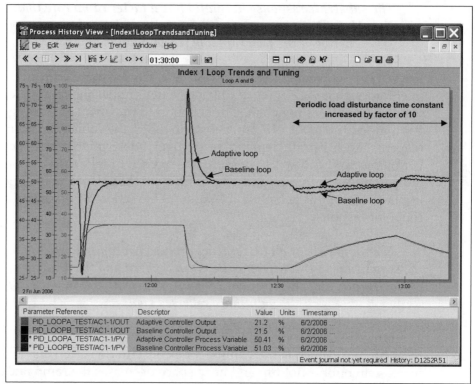

Figure 3-3f. The Effect of Load Disturbance Speed on Process Recovery

important process variables [10]. This opportunity is discussed in chapter 5, on the virtual plant.

The open loop response of these concentrations during the exponential growth and product formation phases is a one-direction integrator. A change in the DO, pH, temperature, and substrate concentration affects the rate by which the amount of biomass and product increases. The biomass and product do not decrease unless the batch has gone too long or has severe problems such that the death rate exceeds the growth rate or the hydrolysis rate (i.e., the rate of consumption of product as a food source by cells) exceeds the product formation rate. The fact that a PID loop could not decrease the biomass or product is problematic. Integral (reset) action cannot be used, and the bias for P-only or PD control would change with batch time. Furthermore, the controller may try to stop the rate of rise. However, if the biomass or product's rate of change, which is the slope of the profile, is the controlled variable, then batch profiles can be directly enforced and the controlled variables (rates of change) can be decreased as well as increased. By translating controlled variables, the proportional mode gives derivative action, the integral model gives proportional action, and the derivative mode gives second derivative (acceleration correction) action. Integral action with its fatal problem is inherently eliminated in terms of the original controlled variables. Also, the integrating process can now be modeled as a self-regulating process. This innovative technique is explored in chapter 4 on model predictive control [11]. The large dead time and time constant caused by slow kinetics and low signal-to-noise ratio make this an excellent candidate for model predictive control.

> *Biomass and product concentration have a one-direction integrating response that poses severe problems for the conventional implementation of PID control.*
>
> *Translating controlled variables from biomass and product concentration to rate of change (slope) offers significant inherent advantages for batch profile control.*

Lambda tuning equations specifically designed for pure integrating processes are available. Appendix C, "Unification of Controller Tuning Relationships," details the use and reduction of these more exact equations into the simpler unified form shown here for the "near" integrating processes of common bioreactor loops. The appendix shows how the more exact equations for "pure" integrating processes provide a maximum controller gain that is about 50 percent larger than what is estimated from equation 3-3i for the low integrator gains seen in bioreactors. Since the actual controller gain used in practice is typically far below this maximum, the more complex equations in appendix C are more useful for deploying in software packages for controller tuning than for understanding or estimating tuning settings.

3-4.   Adaptive Control

Theory and Reality
Process control systems assume a constant linear process. Unfortunately, all process variables and control valves are nonlinear to some degree: the process response to a given change in the controller output changes with batch time, seed culture, and bioreactor operating conditions. The lack of consistency in the process response has significant implications for the process's performance not only in terms of tuning controllers but of recognizing degradations and achieving optimums [12].

Road Maps and Terrain
Consciously or subconsciously, tuning controllers involves a tradeoff between performance and robustness. The controller's ability to tightly control at an operating point is inversely proportional to its ability to weather changes in the plant's behavior without become oscillatory. The operating environment for most loops is stormy, and the last thing you

want is for a control loop to introduce more variability. Consequently, all controllers are detuned (backed off from maximum performance) to some degree to provide a smooth response, despite the inevitable changes in the process dynamics. A PID controller approaches turns cautiously since it doesn't know what lies ahead [12].

PID controllers are backed off from best performance because of the uncertainty of tuning settings.

Controller tuning settings can be computed from a first-order or an integrating-plus-dead-time process model. The changes in the model parameters reveal changes in the cells, process conditions, equipment, final elements, and sensors. The size, direction, and characteristics of these changes can provide a road map and knowledge of the terrain [12].

Changes in the parameters of a dynamic process model identified by an adaptive controller provide insight into changes in the process, equipment, and sensor.

Glimpses and Grimaces
Nearly all of the industrial adaptive controllers presently used in industrial processes require that changes in the process variable be observed over rather a long time and they show the results in terms of new tuning settings. The tuning rules are imbedded and usually unknown. The most commonly used adaptive controller today operates by using pattern recognition and, if it's deemed necessary, it increases the controller gain to induce oscillations. The size of the transients or oscillations and the time required for identification can translate into significant process variability and an adaptation rate that is slower than the rate of change of the process parameters. In fact, most adaptive controllers are playing catch-up even if they have seen the same situation a thousand times before. At best, these controllers provide a snapshot of the current tuning requirements but no real process insight into where the process has been or where it is going. Also, sudden unexplained shifts in the tuning settings or bursts of oscillations reduce the operator's confidence and decrease the likelihood that the controllers will run in the adaptive mode and be used in future applications [11] [12].

Model-free and pattern-recognition adaptive controllers do not offer process knowledge and are playing catch-up.

Watching but Not Waiting
The next generation of adaptive controllers can identify a process model quickly and automatically and provide process model parameters that can be displayed, trended, and diagnosed. Furthermore, these controllers

remember the results for similar conditions, eliminate repetitious identification, and take the initiative [12].

An adaptive controller with these desirable features has been demonstrated in plant tests. The controller can identify the dead time, process gain, and time constant for both manipulated and disturbance variables and save these as a function of a key variable. The user can use the recommended tuning method or elect to choose an alternative method to compute the current tuning settings for the current and memorized conditions. When the key variable indicates that the process has changed, the tuning is then scheduled based on the process model saved in the operating region. The adaptive controller remembers the results from previous excursions and does not wait to recognize old territory. For example, in loops that have nonlinear installed valve characteristics and nonlinear controlled variables, the model and tuning are scheduled based on the controller output and input, respectively. To change dynamics as the batch progresses, the model and tuning are scheduled based on totalized feed. The adaptive controller takes preemptive action based on operating region and uses the opportunity to refine its knowledge of the process model. Changes in these models can flag changes in seed cultures [12].

Adaptive controllers should learn, remember, and utilize knowledge gained from previous batches.

The adaptive controller computes the integrated squared error (ISE) between the model and the process output for changes in each of three model parameters from the last best value. To explore all combinations of three values (low, middle, and high) for three parameters, twenty-seven models are ultimately generated. The correction in each model parameter is interpolated by applying weighting factors that are based on the ISE for each model, normalized to a total ISE for all the models over the period of interest. After the best values are computed for each parameter, they are assigned as the middle values for the next iteration [13]. This model switching with interpolation and recentering has been proved mathematically by the University of California, Santa Barbara, to be equivalent to least square identification and provides an optimum approach to the correct model [14]. The search is actually done sequentially, first for the process gain, then the dead time, and finally the time constant, which reduces the number of models to nine [13]. Figure 3-4a shows the setup of an adaptive controller that identifies the process model for the controlled variable's (ΔCV) response to changes in the controller output (ΔCO) and disturbance variable (ΔDV). This response is then used to compute the feedback controller tuning settings and the feedforward dynamic compensation, respectively [12].

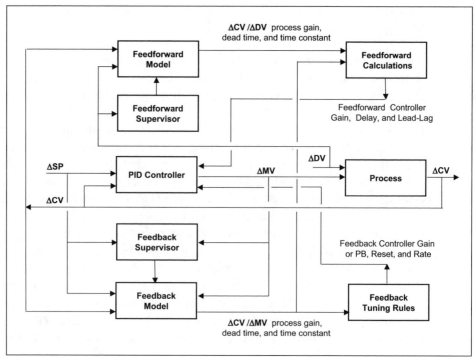

Figure 3-4a. Adaptive Controller Setup Based on Identifying Process Models

The adaptive controller starts in the "Observe" mode, in which it continuously and automatically identifies the process model when it sees changes in the controller's set point, output, or feedforward. The adaptive controller can also be switched to the "Learn" mode, in which it updates the feedback and feedforward process models in each region for which it sees an excursion. The next option is the "Schedule" mode, where the adaptive controller uses the models in each region to change the controller settings. The highest level is the "Adapt" mode, in which the adaptive controller immediately uses any identified improvements [12].

> *An adaptive controller can be run in the "Observe" and "Learn" mode in order to see what it can capture as process knowledge as a function of batch time or variables.*

Back to the Future
This new generation of adaptive controllers allows all PID loops to run in the adaptive mode and provides process model parameters that are saved in a data historian and analyzed for changes in the plant, sensors, and valves. The information on changes in the process model may be directly used to monitor loop performance and to provide more intelligent diagnostics. The models can provide the dynamics for simulations and identify candidates for advanced control techniques. For example, loops

that have large dead times or a one-way integrating response are prime candidates for model predictive control [12].

3-5.   Set-Point Response Optimization

Nothing Says Forever Like Tradition
There are four major practices for starting up a loop that has a large process time constant or slow ramp time compared to the dead time [15]:

Loop Practices for Fast Batch and Startup Response

1. Switch controller to auto with the final set point.

2. Switch controller to auto with an initial set point and then switch to final set point.

3. Put controller in manual or output tracking with final set point, set valve to its normal position, wait, and switch to auto.

4. Put controller in manual or output tracking with final set point, set valve to extreme position, wait, switch valve to normal position, wait, and switch to auto.

Figure 3-5a shows the batch or startup response of a pressure loop that has an integrating response for practices 1, 2, and 4. Practice 3 is not shown because it is not viable for integrating processes. Other practices exist, such as ramping the set point, for unit operations in which it is desirable that the process variable's approach to set point and the output to its final resting value are moderated or that a profile is enforced [15].

The controller output must be positioned beyond the final resting value (balance point) in an integrating process in order to get the CV to move toward set point.

In the first and second practices, the controller output is at its initial value at one end or the other of the output scale (often zero). All methods assume that the pump and block valves have already been started and opened, respectively [15].

In the first practice, if the loop is tuned to minimize variability in the controller output, which is the case for surge volume level control, then the batch phase may time out before the process reaches set point. For example, if the process time constant is 50 minutes and a Lambda factor of five is used, then the closed loop time constant is 250 minutes and the time to reach 98 percent of set point is 1000 minutes (four closed loop time constants). A similar situation exists for slow ramp rates [15].

A dead time that is much faster than the process time constant or ramp rate usually means that a Lambda factor of less than one (i.e., a closed loop time constant or arrest time that is less than the open loop time constant or arrest time) is permissible to achieve stability and is desirable to achieve fast control of these process variables. This is particularly important for the practice 1 because you are relying on reset action to get you set point. All the batch or startup responses in figure 3-5a use a Lambda factor of less than one [15].

If the kick from the proportional mode is negligible, the burden is entirely on the integral mode (reset action), and the approach to set point will be much slower.

In the second practice, the set point is changed from its initial to final value one execution or more after the controller is switched to auto. Note that if you switch the set point within the same execution of the module as the switch of the mode, you will probably end up with the same response as the first practice. In the second batch or startup response, the set point change kicks the output, which gives the process variable a boost on its way to set point. The time to reach set point (rise time) is nearly cut in half, but the settling time is about the same. Since the overshoot is minimal, the rise time might be more important. Also, the controller tuning could be tweaked to reduce settling time [15].

If the set point change is made during the same execution as a mode change, there will probably be no kick from the set point change, despite a β factor greater than zero.

Many of the more astute automation engineers pre-position the controller output by what is called a *head start* or *process action*. For self-regulating loops, the valve position might be set at what was considered to be a normal throttle position or final resting value (FRV), as seen from previous trends when the process variable had settled out at set point. This corresponds to a Lambda factor of one because, if held at this position, it drives the process variable with a time constant that is equal to the process time constant [15].

A Lambda factor of one for self-regulating loops gives an approach to set point at the speed seen in the open loop response for a controller output that is moved to the FRV.

For integrating responses, the process variable won't go anywhere until the valve is positioned beyond its FRV. This leads us to practice 4, in which the valve is set to an extreme position allowable by the process in order to give the fastest approach to set point. Then the brakes are slammed on so the process variable does not run over the set point. The question is, When do you hit the brakes [15]?

The Wait
The plot for the fourth practice shows the response for a technique in which the rate of change is computed from the change in the controlled variable (CV) over a time period long enough to get a good signal-to-noise ratio. The old value of the CV, created by passing the CV through a dead-time block, is subtracted from the new CV. The delta CV is divided by the block dead time to create a rate of change. The rate of change multiplied by the process dead time is then the predicted change in the CV that when added to the new CV is the predicted end point as shown in equation 3-5a. When the end point equals or exceeds the final set point, the controller output is switched from maximum throttle to its FRV. It is held at this FRV for one process dead time and is then released for feedback control. This method compensates for nonlinearities and disturbances that are evident at the time to hit the brakes [15].

The controlled variable's rate of change multiplied by the loop's dead time and added to the old controller's output provides a prediction of the end point.

$$CV_f = [(CV_n - CV_o) / DT] * \tau_d + CV_n \qquad (3\text{-}5a)$$

where:
CV_f =   predicted CV one dead time into the future (%)
CV_n =   new CV (%)
CV_o =   old CV (output of dead time block) (%)
DT =   DT block dead time (sec)
τ_d =   total loop dead time (sec)

If the process dead time is underestimated, the loop overshoots the set point. Therefore, it is important to be generous in the dead time estimate. It is especially important that the dead time not be too short for the zero load integrating process, where the FRV is zero and there is nothing to bring the process variable back to set point [15].

It is better to overestimate the dead time when predicting the end point.

Without Dead Time I Would Be Out of a Job
If the loop dead time is zero, the loop could switch to the FRV when the CV reached set point. Furthermore, the sky is the limit, for the controller gain and feedback action could provide instantaneous correction. My lifestyle is largely the result of dead time.

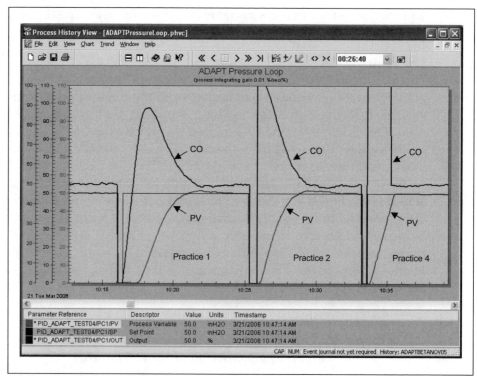

Figure 3-5a. Batch and Startup Performance of an Integrating Loop

A better term than process dead time is *total loop dead time* because there are many sources of dead time outside of the process. The biggest source for slow ramp times and process time constants is the measurement and valve resolution. The time it takes for the change to get through the resolution limit is dead time. The dead time from the measurement and valve resolution are inversely proportional to the rate of change of the process variable and controller output, respectively. Consequently, for closed loop tests the identified dead time depends on the controller tuning and the size of the change in the set point. Fortunately, the changes in valve position are quite large, and the dead time from resolution is minimal for the optimal switching method described in practice 4.

Other sources of instrument dead time include measurement transportation delay, sensor lags, transmitter dampening, analog and digital filters, module execution time, valve dead band, and actuator prestroke dead time.

An adaptive controller can identify the total loop dead time accurately if the trigger points in terms of output changes are large enough. Note that the ultimate proof in the pudding is the output change rather than the set point change. This is because output change includes the effect of tuning and is ultimately what is driving the process. Given the measurement and valve resolution, the adaptive controller with its knowledge of the integrating process gain can correct the observed dead time to give a value that is closer to the output changes associated with the optimal switching [15].

An adaptive controller can also identify the integrating process gain. This can be used with the current ramp rate and the pre-positioned extreme controller output to estimate the FRV, following equation 3-5b. Note that if the extreme output (CO_x) is less than the FRV, the signs of each expression are reversed to get a positive FRV. Of course, limits should be enforced on the calculated value, and it may be desirable to estimate the new FRV by using a portion of the difference between the calculated FRV and the last captured FRV added to the last captured FRV. For primary loops in a cascade control system, the extreme output must match up with the set point limits of the secondary loop and the FRV *is* a set point of the secondary loop. It is necessary to keep the units of the process variable and output consistent with the process integrating gain. If process integrating gain is expressed in %/sec/%, then the process variable and output must both be in % [15].

An adaptive controller can identify the process gain and dead time that are essential for improving the evaluation, performance, and monitoring of optimal switching.

For integrating processes and CO_x > FRV:

$$FRV = CO_x - [(CV_n - CV_o) / DT] / K_i \qquad (3\text{-}5b)$$

where:

FRV	=	final resting value (%)
K_i	=	integrating process gain (%/sec/%)
CO_x	=	controller output at extreme allowed by process (%)
CV_n	=	new CV (%)
CV_o	=	old CV (output of dead time block) (%)
DT	=	DT block dead time (sec)

With a little ingenuity, similar equations can be developed for estimating the FRV of self-regulating processes based on an identified process gain. These equations can be put on line in the observation mode to see how well they estimate the FRV before you actually use the FRV for practice 4. If the FRV is too variable and cannot be accurately captured or calculated, it is best to revert to the second practice. The second practice depends more heavily on the controller's tuning and in particular on the relative amount of proportional and reset action. This is because the tuning is responsible not just for correcting the FRV but for taking the output all the way from its extreme to the FRV. It is important that gain dominates reset action in the approach to set point. Proportional action must kick the output to the allowable extreme and then back it off as the CV approaches set point. This is despite the effect of the reset action, which works to force the output to its limit until the CV crosses set point [15].

The optimal switching technique based on the CV's rate of change is ideally suited for an integrating or ramping process. However, it works well for self-regulating processes in which the fastest possible approach to set point is desired. It also reduces the dependency on tuning since the PID only has to correct for errors in the dead time and the FRV [15].

References

3-1.   McMillan, Gregory K., *A Funny Thing Happened on the Way to the Control Room*. Reprint via ProQuest UMI "Books on Demand", 1989.

3-2.   McMillan, Gregory K., *Good Tuning: A Pocket Guide*. 2d ed. ISA, 2005.

3-3.   McMillan, Gregory K., *Tuning and Control Loop Performance*, 3d ed. ISA, 1992.

3-4.   Åstrom, Karl, and Hägglund, Tore, *Advanced PID Control*. ISA, 2006.

3-5.   McMillan, Gregory K., and Weiner, Stan, "Control Mythology." *Control* (April 2006).

3-6.   Shinskey, F. G., *Feedback Controllers for the Process Industries*. McGraw-Hill, 1994.

3-7.   Blevins, Terrence L., McMillan, Gregory K., Wojsznis, Willy K., and Brown, Michael W., *Advanced Control Unleashed: Plant Performance Management for Optimum Benefits*. ISA, 2003.

3-8.   Skogestad, Sigurd, "Simple Rules for Model Reduction and PID Controller Tuning." *Journal of Process Control* 13 (2003): 291-309.

3-9.   Ziegler, J. G., and Nichols, N. B., "Optimal Settings for Automatic Controllers." *Transactions ASME* 64, no. 11 (1942):759.

3-10.   Wilson, Grant, McMillan, Gregory K., and Boudreau, Michael, "PAT Benefits from the Modeling and Advanced Control of Bioreactors." *Emerson Exchange*, 2005.

3-11.   Trevathan, Vernon L. (editor), *A Guide to the Automation Body of Knowledge*. "Chapter 3 - Continuous Control." Wade, Harold (author), ISA, 2005.

3-12.   McMillan, Gregory K., Sowell, Mark, and Wojsznis, Peter, "The Next Generation – Adaptive Control Takes a Leap Forward." *Chemical Processing*, September 2004.

3-13.   Hespanha, Joao P., and Seborg, Dale E., "Theoretical Analysis of a Class of Multiple Model Interpolation Controllers." Presentation at AIChE Conference, San Francisco, 2003.

3-14.   Wojsznis, Willy K., Blevins, Terrence L., and Wojsznis, Peter. "Adaptive Feedback/Feedforward PID Controller." Presentation at ISA EXPO, Houston 2003.

3-15.   McMillan, Gregory K., "Full Throttle Batch and Startup Response." *Control*, May 2006.

3-16.   Shinskey, F. Greg, "The Power of External Reset Feedback." *Control*, May 2006.

Chapter 4:
Model Predictive Control

Chapter 4

Model Predictive Control

4-1. Introduction

Because Model Predictive Control (MPC) uses an experimental model it can create a future trajectory of the process response based on multiple measured process inputs. Unknown disturbances, noise, limit cycles, random error, and incorrect model parameters result in a bias correction and a shift of the trajectory. Since MPC seeks to minimize the squared error of the trajectory over the time horizon, the short-term effects of unknowns and erratic signals are minimized. In contrast, a PID only knows what it sees as a change or rate of change for one process feedback measurement in the current scan. Additionally, MPC can simultaneously manipulate multiple variables, whereas a PID control block is restricted to one controller output and one feedback measurement. Consequently, PID must use various downstream blocks, such as the "Splitter" block for split-range control of multiple manipulated variables and the "Control Selector" block for override control of multiple controlled variables. With all of these PID techniques, a single controlled variable is matched up with a single manipulated variable at any given time with no inherent knowledge of the dynamics of the pairing. PID uses sequential pairing whereas MPC offers simultaneous manipulation of multiple process inputs for control of multiple process outputs. Finally, MPC is inherently better at optimization because it can predict future violations of constraints; has built-in features for maximization, minimization, and setting priorities; has a tuning adjustment to smooth out the optimization; and has hooks to a linear program.

Section 4-2 of this chapter addresses how process gains and an inaccurate model dead time prevent MPC from doing its job. MPC becomes more sensitive to an inaccurate dead time if the process is dominated by dead time or the process gains indicate that the process inputs are not actually independent variables, that the process outputs do not strictly depend upon the process inputs, or that some process inputs have too great or too little of an effect. This section discusses the performance advantages of MPC in terms of its functional capability and concludes with a summary of its advantages.

Section 4-3 describes the setup, tuning, and performance of MPC for the simultaneous manipulation of multiple process inputs. In particular, the section details MPC's ability to optimize process inputs while rejecting load disturbances and eliminating split-ranged control. A standard MPC block has been offered in distributed control systems for more than a decade.

Section 4-4 begins with an overview of the optimum relationships between biomass growth rate and process conditions, such as temperature, pH, DO, and substrate concentration. The section alerts the user to a potential change in the sign of the process gain that necessitates a change in controller action. The section also shows how translating controlled variables is essential for optimization of the batch profiles of biomass and product concentrations. It ends with an example of the MPC models and setup.

Learning Objectives

A. Understand how the MPC algorithm gains knowledge of the future from the past.

B. Recognize the implications of MPC's long-term view versus the PID's short-term view.

C. Appreciate MPC's ability to deal with erratic responses and interaction.

D. Know how to check whether an MPC solution is viable.

E. Be able to summarize the advantages of MPC.

F. Know how to set up an MPC block for optimizing a process input.

G. Be able to set up an MPC block to replace split-range control.

H. Recognize the nature of the optimization problem.

I. Determine the requirements for a change in control action for optimization.

J. Understand the translation of controlled variables needed for profile control.

K. Be aware of the type of models and MPC setup used for profile control.

4-2.   Capabilities and Limitations

Model predictive control (MPC) uses incremental models of the process in which the change in controlled or constraint variables is predicted from a history of changes in the manipulated variables, called "moves," and changes in the disturbance variables. The initial values of the controlled, constraint, manipulated, and disturbance variables are set to match those in the plant when MPC is in the initialization mode. MPC can be run in manual and the trajectories monitored. If the trajectories drastically change or flip-flop from one execution to another, then the models need to be improved or the time to steady state increased. Often, the significant delay and slowness in the observed response helps operations and process technology personnel appreciate the implications of the process dynamics. While in manual, a back-calculated signal from the downstream function blocks is used. This makes it possible for the manipulated variables to track the appropriate set points so MPC is ready for a bumpless transition to automatic. When a new MPC application is commissioned, the MPC block maximum and minimum output limits are narrowed until experience and confidence is gained with the model accuracy and optimization strategy [1] [2].

The trend in the future response to current disturbances and past moves helps operations and process technology personnel appreciate the implications of process dynamics.

A simplified review of MPC's functionality will help us gain the understanding essential to our later discussion of MPC's capabilities and limitations. Figure 4-2a shows the response of a change in the controlled variable to a step change in each of two manipulated variables at time zero. If the step change in the manipulated variables were twice as large, the individual responses of the controlled variable would be predicted to be twice as large. The bottom plot shows the linear combination of the two responses. Nonlinearities and interdependencies can cause this principle of linear superposition of responses to be inaccurate. Any errors in the modeled-versus-actual process response shows up as an error between the predicted value and actual valve of the controlled variable, as shown in the upper plot of figure 4-2b. A portion of this error is then used to bias the process vector, as shown in the middle plot. The MPC algorithm then effectively calculates a series of moves in the manipulated variables that will provide a control vector that is the mirror image of the process vector about the set point. This is shown in the bottom plot of figure 4-2b. MPC has a built-in knowledge of the combined inverse response where the initial reaction of the CV is in the opposite direction of the final change in the CV. If there are no nonlinearities, load upsets, or model mismatch, the predicted response and its mirror image should cancel out, with the controlled variable ending up at its set point. How quickly the controlled variable reaches set point depends on the process dynamics, "penalty on move" (PM, penalty for changes to a manipulated variable), move size limit, and number of moves set for the manipulated variables [1] [2].

MPC can handle interactions better than PID because it can simultaneously manipulate multiple process inputs so as to control the predicted response of multiple process outputs to the combined effects of disturbances and previous process inputs.

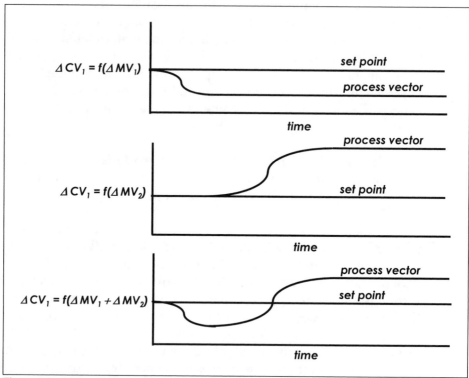

Figure 4-2a. Linear Superposition of MPC Response

The first move in the manipulated variable is actually the linear summation of all the first moves based on controlled, disturbance, and constraint variables. Only the first move is executed because the whole algorithm is revaluated in the next controller execution.

Figure 4-2c shows how MPC has a view of the trajectory of each controlled and constraint variable based on changes in the manipulated and disturbance variables. The prediction horizon is the time to steady state, which is approximately the time to 98 percent of the final value (T_{98}), as defined by equation 2-3a in chapter 2.

In contrast, a PID controller only sees the current value and the rate of change of its controlled variable. Adding a dead-time compensator only extends the controller's view of the future to the end of the dead time, as shown in figure 4-2c. The derivative (rate) mode provides some

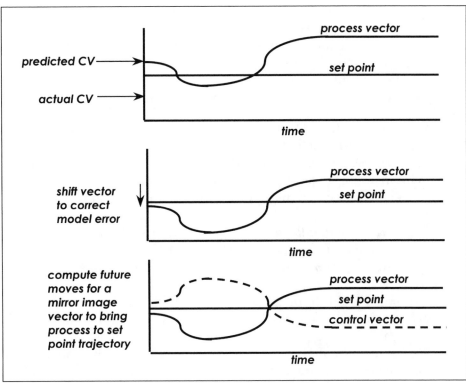

Figure 4-2b. Shift of Process Vector and Mirror Image Control Vector

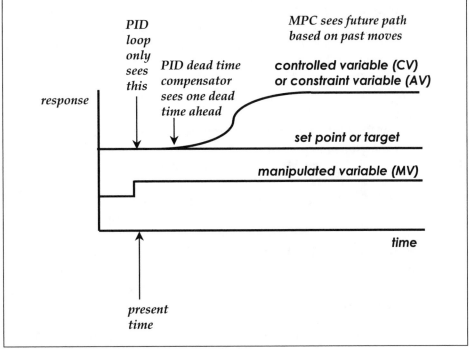

Figure 4-2c. MPC and PID Views of the Process Response

anticipation based on the controlled variable's slope of response, and the proportional (gain) mode gives an immediate and abrupt action. Feedforward and decoupling can be added, but the addition of these signals to the controller output is based on current values, has no projected future effect, and is designed to achieve the goal of returning a single controlled variable back to its set point [1] [2].

MPC takes a long-term view of the process response, whereas the PID reacts to the immediate situation. Consequently, abrupt, erratic, or nonrepresentative changes, such as steps or a scatter of readings from a chromatograph, resolution limits, process and measurement noise, and inverse response are more problematic for a PID controller.

MPC can handle noise, resolution limits, steps, random error, and erratic response better than PID because MPC has a longer-term view of the process response.

Tuning an MPC application can often be reduced to a single tuning factor called "penalty on move" (PM), also known as "move suppression," which sets the degree of transfer of variability from the controlled variable to the manipulated variable. The PM combined with a move size limit offered by MPC prevents large and fast changes in the manipulated variables.

The performance advantages of MPC that directly relate to economic benefits are its inherent capability to meet process objectives, manipulate multiple outputs, and deal with interactions. Since it also computes the trajectories of constrained variables and has built-in capabilities for suppressing moves and maximizing or minimizing a manipulated variable, it is well suited to multivariable control and optimization. Move suppression provides the smoothness desired for optimization.

MPC's built-in ability to handle interactions, predict future violations of constraints, and smoothly maximize or minimize an MV is important for optimization.

As with the PID controller, a tradeoff exists between an MPC's performance (minimum peak and integrated error in the controlled variable) and its robustness (maximum allowable unknown change in the process gain, dead time, or time constant). Higher performance corresponds to lower robustness. However, the direction of the dead-time change that most destabilizes MPC is the opposite of the change that destabilizes a PID [1].

MPC's robustness is more sensitive to a decrease than an increase in dead time. This is especially true in dead-time-dominant processes, where the total dead time is much greater than the process time constant. A decrease in plant dead time can rapidly lead to growing oscillations that are much faster than the ultimate period. A large increase in dead time shows up as much slower oscillations, with a superimposed high-frequency limit cycle. In contrast, a PID can become unstable for an increase in total loop dead time. A decrease in dead time for a PID translates to lost opportunity in performance because the PID becomes less aggressive than is possible in terms of its reset and gain settings [1] [2].

> *Unlike PID control, MPC is oscillatory for an actual plant dead time that is either smaller or larger than the identified dead time. MPC is also more sensitive to and destabilized by a decrease in plant dead time in dead-time-dominant processes.*

PID controllers thrive on the steady and gradual response of a process time constant or process integrating gain that is much slower than the total dead time. If the PID has a single manipulated variable and no interactions, PID control can achieve a level of fast, unmeasured load disturbance rejection that is difficult to beat. However, for bioreactors the disturbances are so slow and the PID is so detuned that moving to MPC with a good model may improve control even in cases where the PID normally excels.

For measured disturbances, MPC generally has a better dynamic disturbance model than a PID controller with feedforward control. This is primarily because it is difficult for the user to properly identify the feedforward lead-lag times for PID. Often, the feedforward dynamic compensation for PID controllers is omitted or adjusted by trial and error.

For constraints, MPC anticipates a future violation by looking at the final value of a trajectory versus the limit. MPC can simultaneously handle multiple constraints. PID override controllers handle constraints one at a time through the low or high signal selection of PID controller outputs.

For interactions, MPC is much better than a PID controller. The decoupling signals sometimes added to a PID are generally exclusively computed based on steady-state gains. This PID decoupling ignores the effects of the relative speed of loops on interaction. However, the benefits of MPC over detuned or decoupled PID controllers deteriorate as the condition number of the matrix increases.

The steady-state gains of the 2x2 matrix in equation 4-2a show that each manipulated variable has about the same effect on the controlled variables. The inputs to the process are linearly related. The determinant is nearly zero and provides a warning that MPC is not a viable solution [1] [2].

$$\begin{bmatrix} \Delta CV_1 \\ \Delta CV_2 \end{bmatrix} = \begin{bmatrix} 4.1 & 6.0 \\ 4.4 & 6.2 \end{bmatrix} * \begin{bmatrix} \Delta MV_1 \\ \Delta MV_2 \end{bmatrix} \qquad (4\text{-}2a)$$

The steady-state gains of a controlled variable for each manipulated variable in equation 4-2b are not equal but exhibit a ratio. The outputs of the process are linearly related. Such systems are called "stiff" because the controlled variables move together. The system lacks the flexibility to move them independently to achieve their respective set points. Again, the determinant is nearly zero and provides a warning that MPC is not a viable solution [1] [2].

$$\begin{bmatrix} \Delta CV_1 \\ \Delta CV_2 \end{bmatrix} = \begin{bmatrix} 4.1 & 6.0 \\ 2.2 & 3.1 \end{bmatrix} * \begin{bmatrix} \Delta MV_1 \\ \Delta MV_2 \end{bmatrix} \qquad (4\text{-}2b)$$

In equation 4-2c, the steady-state gains for the first manipulated variable (MV_1) are several orders of magnitude larger than for second manipulated variable (MV_2). Essentially, there is just one manipulated variable MV_1 since the effect of MV_2 is negligible in comparison. Unfortunately, the determinant is 0.9, which is far enough above zero to provide a false sense of security. The condition number of the matrix computed from the singular values of the matrix provides a more universal indication of a potential problem than either the determinant or relative gain matrix. A higher condition number indicates a larger problem. At some point, the condition number becomes too high. For equation 4-2c, the condition number exceeds 10,000 [1] [2].

$$\begin{bmatrix} \Delta CV_1 \\ \Delta CV_2 \end{bmatrix} = \begin{bmatrix} 1 & 0.001 \\ 100 & 1 \end{bmatrix} * \begin{bmatrix} \Delta MV_1 \\ \Delta MV_2 \end{bmatrix} \qquad (4\text{-}2c)$$

The matrix condition number provides a numerical assessment of potential MPC performance problems.

The matrix can be visually inspected for indications of possible MPC performance problems by looking for gains in a column that has the same sign and size, gains that differ by an order of magnitude or more, and gains in a row that are a ratio of gains in another row. Very high process gains may cause the change in the MV to be too close to a control valve's dead band and resolution limits, and very low process gains may cause an MV to hit its output limit [1] [2].

The desired relationship for MPC is for process outputs to depend only upon measured process inputs that are independent variables. If the process inputs affect each other they are not independent. If the process outputs affect each other then they are not solely dependent on the process inputs. If the process gains associated with an MV are more than an order of magnitude greater or smaller than process gains of other MV, then the effect of process inputs cannot be sufficiently balanced for control.

MPC performance relies on process inputs that are independent of each other, process outputs that depend on process inputs rather than other process outputs, and the set of process gains for each MV having a comparable magnitude.

MPC software packages compute the condition number after the models have been identified for the variables included in the control matrix [3]. The setup screen gives a relative index of the importance of each manipulated variable and computes the condition number for the controlled and manipulated variables selected. The user can move variables in and out of the control matrix and see the effects on the condition number [4].

A condition number above 100 indicates a serious degradation in both performance and robustness. MPC may work well enough up to a point if the "penalty on move" is increased and the changes in set points are coordinated. However, the best solution is to come up with different controlled or manipulated variables. The problems associated with a condition number that is greater than 1000 cannot be sufficiently mitigated by tuning. One innovative possibility is to use principal component analysis, as described in chapter 8, to provide new independent latent variables to replace existing collinear controlled variables (i.e., controlled variables with linear interrelationships).

Some MPC software help guide users in selecting manipulated and controlled variables in the control matrix in order to reduce the condition number.

To summarize, if the condition number is not too high, then MPC is useful for the situations listed in the following:

Situations Where MPC Can Be Beneficial

1. Process and measurement noise
2. Erratic or stepped measurement response
3. Inverse response
4. Large dead times

5. Move size limits
6. Measured disturbances
7. Multiple manipulated variables
8. Interactions
9. Constraints
10. Optimization

The remaining sections of this chapter will focus on applying MPC to situations 7 through 10.

The following business and practical issues need to be addressed in order for MPC to be applied to industrial biopharmaceutical processes [5]:

1. Today, the vast majority of basic control loops in major pharmaceutical companies are satisfactorily handled as single input single output loops.

2. Perturbing the system to generate the model, such as changing temperature and pH in a running fermentation for the sake of identifying a model, is not something that most plants want to do. This is because even small excursions from set point, especially involving temperature and pH, can negatively affect the culture.

3. The perception is that MPC is so difficult to set up that an outside firm or consultant is typically needed.

4. The lack of research papers in the literature showing how MPC in bioprocesses has provided a significant financial benefit.

5. The lack of familiarity with MPC among automation and process control support engineers.

6. The extra cost of MPC software. With many control systems, PID is standard off-the-shelf functionality that comes with the system, but MPC is an often somewhat expensive add-on software package.

7. The perception that MPC was developed for and is best suited for steady-state continuous processes, such as those found in the petrochemical industry. Most pharmaceutical processes are non-steady-state batch processes with several step-to-step transitions.

8. Discomfort among development and manufacturing engineers, as well as quality control folks, over the formal validation of MPC applications, which is a cGMP requirement for systems making medicines for human consumption. Specifically, how does the user make regulatory professionals comfortable that a controller is robust enough to handle contingencies and disturbances that it had not been exposed to when the model was

created and when the nature and structure of the model
equations are not disclosed by the MPC supplier?

The benefits from optimization can provide the motivation to advance
beyond single input and output control noted in item (1) to multivariable
control. Solving the remaining issues is facilitated by using relatively low-
cost and easy-to-use software for advanced control algorithms that has
been developed and prototyped in a virtual plant and demonstrated in
bench-top and pilot-plant systems with industrial control systems. Item 8
is not as big an issue as it used to be in that proprietary algorithms have
been validated for use (e.g., freeze thaw skids algorithms).

4-3.   Multiple Manipulated Variables

The relative precision, cost, and speed of each MV depend on the process
and the equipment, piping, and automation system design. For this first
case involving situation 7, we consider the control of dissolved oxygen
(DO) by manipulating air flow and agitator speed for a viscous bacterial
cell culture where the relative cost of agitation power is greater than air
flow. Several examples have demonstrated the successful use of MPC to
control dissolved oxygen by manipulating multiple variables. In one case,
an MPC application controlled the DO in a bioreactor by manipulating air
flow, agitation speed, and vessel pressure [6]. In another case, an MPC
application for a microbial fermentation controlled DO by manipulating
air, oxygen, and nitrogen to the headspace and oxygen sparge to the broth
and it controlled pH by manipulating carbon dioxide to the headspace
and a sodium carbonate solution to the broth [7].

High- and Low-Cost MV
For maximum process efficiency, it is important to reject unmeasured load
disturbances and achieve a smooth optimization. In some cases, the
manipulated variable (MV) with the fastest effect on the process is the
most expensive. This section considers the use of MPC with one high-cost
fast MV and one low-cost slow MV to achieve the dual objective of load
rejection and optimization for a single controlled variable (CV). Many
ways to set up the MPC application are available, but one method that has
been demonstrated to work well is to configure the MPC block for
maximization of the low-cost MV and make a simple tuning adjustment to
reduce the importance of the optimization compared to load rejection [8].

*In general, it is best to optimize the slow MV and keep the fast
MV free for rapid response to disturbances and set points.*

MPC can be configured and tuned to maintain a critical CV (e.g., dissolved
oxygen) at its target and to maximize the low-cost slow MV (air flow) set

point as an optimization variable. The best load and set-point response for the critical CV is obtained through a short-term tradeoff in efficiency: reducing the "penalty on error" (PE) for the optimization variable rather than increasing the "penalty on move" (PM) of either MV. When riding the low-cost MV maximum set point, such as the air flow that can cause foaming, or when riding the high-cost MV low set-point limit, such as the air flow needed to maintain a minimum number of bubbles, this adjustment of the PE rather than the PM lets both the slow and fast MV move to improve the load and set-point response of the critical CV. Only the response of the optimization variable is slowed down. This is consistent with the general theme that disturbance rejection (regulatory control) should be faster than optimization. Figure 4-3a shows the general setup of a 2x2 MPC with a critical controlled variable that has a normal PE and an optimization variable that has a lowered PE [8].

Reducing the "Penalty on Error" (PE) of the optimization variable makes regulatory control more important than optimization.

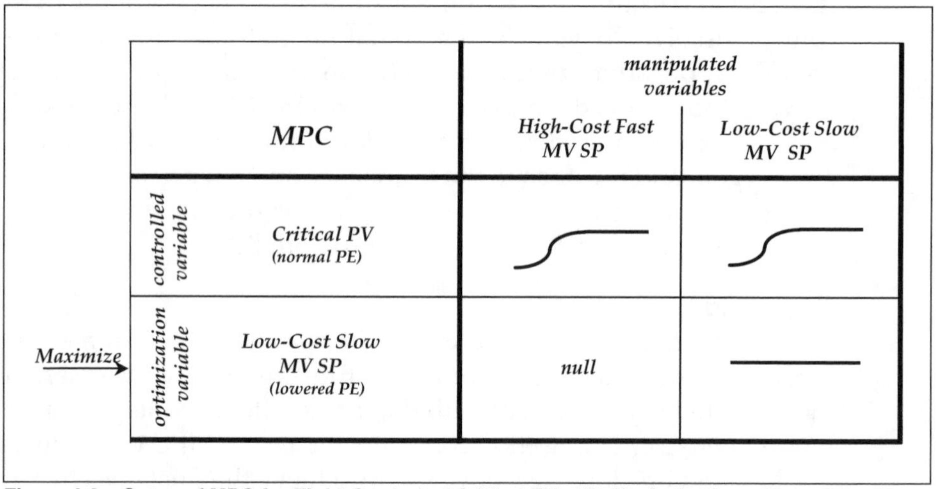

MPC		manipulated variables	
		High-Cost Fast MV SP	Low-Cost Slow MV SP
controlled variable	Critical PV (normal PE)	∫	∫
optimization variable	Low-Cost Slow MV SP (lowered PE)	null	——

Maximize →

Figure 4-3a. Setup of MPC for Tight Control of Critical PV with Maximization of Low-Cost Slow MV

The optimization could also have been slowed down by increasing the penalty on move (PM) on the low-cost slow MV. However, a large PM forces the high-cost fast MV to take most of the correction, particularly if you are riding the low-cost MV maximum set point. Additionally, if you are riding the high-cost MV low set-point limit and the upset or set-point change is in the direction that requires a decrease in this high-cost MV, then the large PM makes the load and set-point response noticeably slower since the response depends totally on the slow MV with its high PM [8].

Leaving the "penalty on move" (PM) of the manipulated variables at the values that are best for performance and robustness allows all of the MV to help with regulatory control.

The test results shown in figure 4-3b are for a process where the dead time is about as large as the time constant. This causes the peak error to be almost the full-load upset because of the high degree of dead time. The improvement is mostly seen in the integrated absolute error (IAE). For processes where the time constant is much larger than the dead time, the reduction in peak error and IAE would be greater [8].

The trend chart in figure 4-3b shows the response for load disturbances and set point changes in both directions. In the first operating mode, the low-cost MV is riding its maximum set point, which leaves the fast cost MV above its low set-point limit, and thus free to respond in the negative as well as a positive direction. In the second operating mode, the maximum for the low-cost MV has been increased to the point where it is no longer achievable, which drives the high-cost MV to its low set-point limit. The succession of load upsets can be so frequent that the high-cost MV doesn't always ride its hard low limit, which is the case shown in figure 4-3b. The test sequence also displays the slow response of the optimization variable to large changes in its maximum set point. It is important for operators to realize that the optimization is purposely designed to be slow and smooth, and in the long term it does a great job [8].

The optimization algorithm has a tight grip on the optimization variable. A large decrease in the PE helped free up the associated MV to move when it was riding its maximum set point. Setting the set point for maximizing the low-cost MV at a level higher than is obtainable freed up this slow MV for upsets and set-point changes that demanded an increase in the MV. However, a higher-than-obtainable slow MV maximum forces the fast MV to its low set-point limit where it is unavailable for corrections that require a decrease in the fast MV [8].

The alternate strategy of minimizing the high-cost fast MV is not advisable for best load rejection because it makes it difficult to free up this fast MV sufficiently once it becomes an optimization variable. However, this might be a viable alternative if it is desirable that the minimum for the fast MV be a soft limit via an economic objective rather than a hard limit via a low set-point limit and if it is sufficient to turn over most of the load rejection to the slow MV. Again, the PE must be reduced for the optimization variable [8].

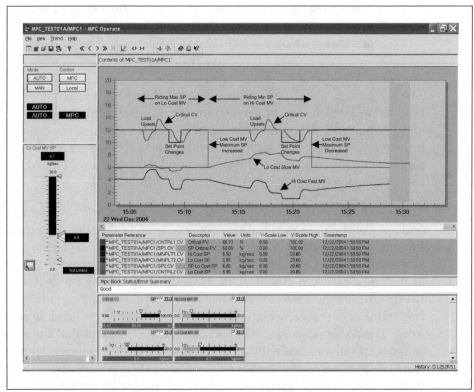

Figure 4-3b. Response of High- and Low-Cost MV MPC to Disturbances and Set Points

Coarse and Fine MV

Two or more manipulated variables are often used to extend the range of loads that loops can handle. Typically, the variables are split-ranged, where the variable with smallest effect (low-gain MV) is throttled for low loads (low controller outputs) and the variable with the largest effect (high-gain MV) is throttled for high loads (high controller outputs). A standard Fieldbus "splitter" block is used on the controller output to split a single controller output into set points for analog outputs. Ideally, to help linearize the loop the split-range point or point of transition from the variables that have small to large effects is based on process gain. For example, if the high-gain MV has nine times the effect of the low-gain MV, then the split-range point should be 10 percent so the low-gain MV and high-gain MV move from minimum to maximum values for a controller output of 0-10% and 10-100%, respectively. However, this practice of compensating for gain by selecting split-range point is uncommon. For the split-ranging of two manipulated variables, the split-range point is traditionally set at 50 percent even though "splitter" blocks, position read-back, speed measurements, and secondary loops have eliminated both special calibrations and operator misunderstandings over actual valve position or drive speed [9].

Besides the process gain, the process dead time and time constant is generally different for each split-ranged output. If there is no compensation for the changes in process dynamics with each manipulated variable, then the controller tuning must be compromised to ensure stability for the worst-case MV.

The controller tuning is compromised to deal with the changes in process dynamics that are associated with split-ranged output.

Unfortunately, the high-gain MV resolution limit, which is expressed in terms of percentage of full scale, will have a bigger effect on the process because of its higher process gain. Also, if the high-gain MV is a rotary valve, the percent resolution limit is bigger to begin with than it is for sliding stem valves or variable speed drives [10]. From here on, the high- and low-gain MV will be referred to as the "coarse" and "fine" (trim) MV, respectively.

Most loops exhibit a "limit cycle" across the split-range point from the gain discontinuity caused by the different MV, the rangeability limits of valves and drives, and the additional friction from both the seating of the plug in sliding stem valves and the sealing of the ball or disc in rotary valves [9].

Most control loops oscillate around the split-range point because of discontinuities, rangeability limits, and stick slip at the split-range point.

To address these issues, "integral only" (I-only) controllers are sometimes used to eliminate the split-ranging. Since these controllers are normally implemented for big and small valves, they are called "valve position" controllers. However, the name "valve position" causes confusion between these new controllers in the control room and the traditional valve positioners mounted on the valves and also unnecessarily excludes other final elements, such as variable-speed drives.

The original process PID controller output now goes to just the trim MV. An "integral only" controller is added with the fine MV signal as its CV; a mid-range value, such as 50 percent, as its set point; and an output that only goes to the coarse MV. The "integral only" controller reset time is set larger than five times the product of the PID controller gain and reset time settings in order to make the interaction between the controllers negligible. Unfortunately, this detuning of the "integral only" controller is too slow to handle disturbances and causes a slow limit cycle from the resolution limit of the coarse MV [9]. Also, a study of the application of dual input (MV) temperature and pressure processes shows that the tuning of these controllers is much more critical than was suspected [11].

Tuning "integral only" controllers to minimize interactions but still ensure responsive manipulation of the coarse MV is critical but difficult.

MPC can be configured to simultaneously manipulate a coarse and fine MV, eliminating the problems inherent in split-ranged and "integral only" controllers. This provides the precision of control offered by the fine MV with the range of control possible from combining fine and coarse MV. Since MPC has the process gain and dynamic response of each MV built in to its model of the process, MPC knows how much to move each MV and inherently eliminate the interactions from the multiple process inputs.

MPC can provide simultaneous fast fine and coarse MV control from its knowledge of the process dynamics for each MV.

The MPC block in figure 4-3c has two manipulated variables (fine MV and coarse MV). These are process inputs, one critical controlled variable (e.g., DO or pH), which is a process output, and one optimization variable (fine MV). The MPC optimization mode selected is "Observe Limit," which will push the fine MV from either direction toward its set point (e.g., 50%). The "penalty on error" for the optimization variable is decreased from 1.0 to 0.1 to increase the relative importance of the control of the critical variable. The tuning setting of the "penalty on move" (PM) computed by MPC was used because the PM was correctly initialized to be much larger (PM=20) for the coarse MV than for the fine MV (PM=5). Using this setting ensures that the coarse MV does not move too fast. If MPC does not automatically tune the PM correctly, the user must manually increase the PM of the coarse MV [9] [10].

The PM for optimizing the fine MV at mid range is reduced to make this optimization less important than regulatory control.

The MPC application could have been set up to optimize the coarse MV to a minimum. However, the set point (target) for this minimization would need to be calculated based on the installed characteristics and capacities in order to keep the fine MV from running near its output limit. This complication is unnecessary with the MPC setup shown in figure 4-3c, which works to ensure the fine MV is fully available by pushing it to the middle of its range.

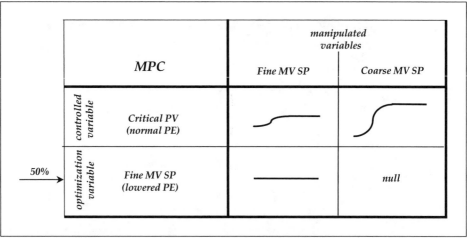

Figure 4-3c. Setup of MPC for Rapid Simultaneous Manipulation of Dual Process Inputs

The trend chart in figure 4-3d shows the response for large steps in load and set points for both the critical PV and the desired optimum position for the fine (trim) MV. Notice that for the successive load upsets, the coarse MV moves rather quickly to a new position, which enables the fine MV to return to its optimum. The load rejection is smooth and fast. Similarly, for a change in the set point of the critical PV from 50 percent to 70 percent, the coarse MV moves to take care of the long-term need at the new set point. Finally, a change of the optimum set point of the fine MV from 50 percent to 60 percent shows only a small bump to the process. Normally, the operator would not be changing the optimum, but a set-point filter could be added for this optimization variable so as to prevent even a small bump [9] [10].

The simultaneous throttling of both valves by MPC will reduce the limit cycle from the coarse MV resolution limitations. If the "Maximum MV Rate" parameter in the MPC setup is written to zero when a separate filtered value of the "critical PV" stays within a band around its set point, then the limit cycle from the coarse MV is eliminated. The width of the band can be calculated as the resolution limit multiplied by the process gain in the MPC model for the coarse MV [9] [10].

> *The limits cycle from the resolution limit of the coarse MV can be eliminated by setting its "Max MV Rate" to zero when the critical PV is close to its set point.*

For those applications in which the manipulated variables have an opposite effect on the process—such as steam and coolant, oxygen and nitrogen, and acid and base reagents—a different setup of the model predictive controller is needed to keep both manipulated variables from being fed at the same time and wasting energy or raw materials [9] [10].

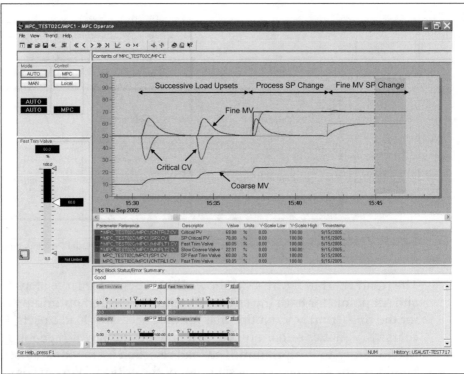

Figure 4-3d. Response of Fine and Coarse MV MPC to Disturbances and Set Points

4-4.   Optimization

The optimum values of the cell environment, such as temperature or pH, depends on process, product, and in some cases the batch phase. Chapter 6 describes the equations for the kinetics employed in first-principle models to quantify the effects of operating conditions on biomass growth rate and product formation rate. Chapter 5 describes how model predictive control and neural networks can be used to adapt kinetic parameters and expressions to changing batch conditions and seed cultures.

The plot of biomass growth rate versus temperature in figure 4-4a has a rather sharp peak. As the temperature approaches the optimum, the growth rate may double for every 10 degrees Centigrade per an Arrhenius relationship over a narrow range. Microbiologists responsible for defining the parameters do an experimental design involving several different temperatures and then a standard curve fit to the resulting experimental results. When the temperature gets high enough to start denaturing proteins and killing cells, the rate of kill is such that a 10 degree C change in temperature results in an order-of-magnitude increase in kill rate. So there are lots of different activities occurring within cells at different

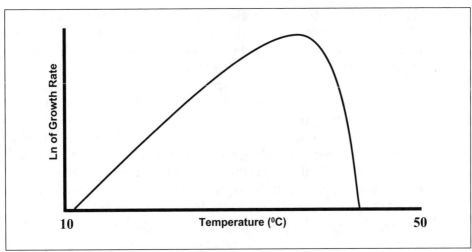

Figure 4-4a. Temperature Effect on Growth Rate

temperature dependent rates. The location of the peak may shift and the slope change for product formation rate [12].

> *There is a sharp drop in growth rate at temperatures greater than the optimum.*

The plot of biomass growth rate versus pH in figure 4-4b has a more symmetrical shape. It may be possible to widen the width of the optimum for some cells by adapting the culture such as by making small changes in the pH of the seed culture prior to its transfer to the bioreactor [13].

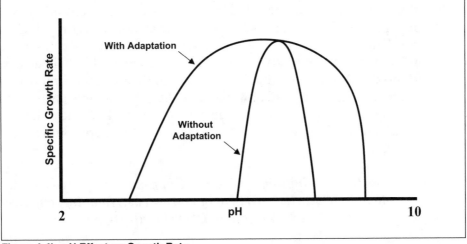

Figure 4-4b. pH Effect on Growth Rate

As with temperature, the optimum pH for product formation rate may be different than the optimum pH for growth rate. Many microorganisms can maintain the intracellular pH to some extent for changes in the pH of their environment. However, the shifts in the broth pH caused by the change in ammonium ions from metabolism of nitrogen sources, the production of organic acids and bases, the utilization of amino acids, and the evolution or feed of carbon dioxide makes pH control important [14]. One major pharmaceutical company's experience in making 20 different products from bioprocesses (including ones made with bacteria, fungi, yeast, and mammalian cell cultures) is that tight pH control is very important and all the optimum pH set points are between 6 and 8. A drift of more than a very few tenths of a pH unit away from set point has negative consequences for the fermentation [15]. Ethanol processes generally run at a lower pH.

pH is important for proper cell metabolism.

The plot of biomass growth rate versus dissolved oxygen in figure 4-4c shows a steep decline below a critical dissolved oxygen concentration (below 0.1 mg/liter). This is because the cells for aerobic processes are deprived of oxygen. High viscosities may decrease mass transfer rate so that higher aeration rate or agitation rate may be needed to achieve the critical concentration. Fermentations that are filamentous in nature (i.e., tangled strands of cell chains that look a little like spaghetti when viewed in a microscope) have tangled webs that grow in size. The broth viscosity will increase (perhaps up over 100 cp), and it may become more difficult for oxygen to get to the cells in the middle of these tangled masses of cells. In this case, it would be prudent to raise the DO set point. Between 0.2 and 0.8 mg/liter, the growth rate is rather constant. Concentrations above 0.8 mg/liter can be toxic, possibly because free radicals form, which accelerates cell aging and even cell death [15]. Foaming can sometimes be related to DO, especially when aeration is increased to help achieve the desired DO (foam creation is a function of the nutrient composition of the broth, aeration rate, and agitation rate) [15]. Foaming is particularly problematic when antifoam agents cannot be added. The cost of air, agitation, and oxygen means there is an economic optimum not shown in figure 4-4c.

Most discussions and plots of the optimum dissolved oxygen do not include the effects of operating cost, foaming, and free radicals at high DO concentrations.

The plot of substrate saturation and inhibition effect on biomass growth in figure 4-4d has a steep decline at low and high concentrations. Not shown is that at low biomass concentrations, such as might occur in the lag phase,

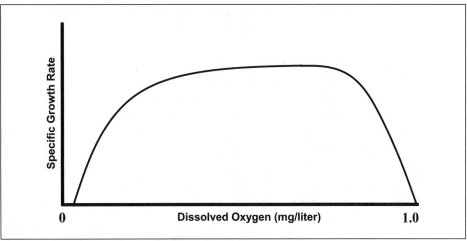

Figure 4-4c. Dissolved Oxygen Effect on Growth Rate

the effect is negligible. Typically, there is no substrate inhibition until the substrate concentration gets to some specific value, with 20-25 g/liter being typical for some glucose fermentations [15]. Therefore, a fermenter is operated at a high substrate concentration in order to maximize specific growth rate in accordance with the kinetic equations presented in chapter 6. However, the concentration is not so high as to cause substrate inhibition to become important. Then, near the end of the fermentation, the substrate concentration is typically lowered to avoid throwing away a lot of unused substrate when harvesting the fermentation. However, achieving substrate concentrations close to zero near the end is difficult, may increase the consumption rate of the product as food (hydrolysis rate), and may starve some cells as a result of localized depletion of food from the low agitation rates used for animal cell cultures.

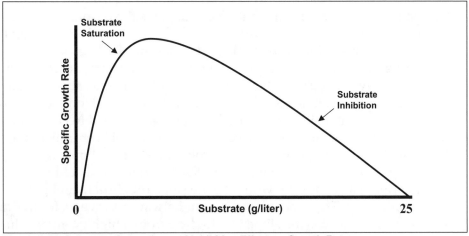

Figure 4-4d. Substrate Saturation and Inhibition Effect on Growth Rate

The typical optimum profile for substrate concentration shows
a decline near the end of the batch to improve yield.

Although the plots shown in figure 4-4a through 4-4d are for biomass
growth rate, similar plots may exist for product formation rates as a
function of broth operating conditions (temperature, pH, DO, and
substrate). In some processes, such as ethanol, there is a product
inhibition, in which the product formation rate goes to zero when the
product concentration reaches a critical value (e.g., 65 g/liter of ethanol). If
the kinetic parameters are known, the plots for biomass growth and
product formation rates can be generated by plotting the results for the
kinetic expressions for each broth operating condition of interest.

In figures 4-4a through 4-4d, any change in sign of the slope corresponds
to a change in the sign of the process gain if the process variable on the
horizontal axis was manipulated to optimize the biomass growth rate. The
controller action must be the opposite of the process action or the
controller output will walk to its output limit. For example, on the plot
biomass versus pH, at low pH the slope ($\Delta G/\Delta pH$) and hence the process
gain is positive (direct action). The controller must have reverse action so
the controller reacts to a decrease in growth rate by increasing the pH.
However, at high pH the slope and process gain is negative (reverse
action). The controller must now have direct action. The solution is to
restrict the controller to working on one side or the other of the optimum.

The change of sign in the slope of the process variable, such as
pH, as it goes from one side to the other of the optimum
requires a change in the controller action.

The actual slope can be computed over a time interval and monitored for
sign if the biomass growth rate can be computed from an online biomass
analyzer or by a virtual plant. The past pH is the output of a dead time
block whose input is the current pH and whose dead time is equal to the
time interval, which is the time to steady state. The change in pH is the
current pH minus the past pH. The change in growth rate is computed the
same way. Of course, other process conditions can be affecting growth
rate, so the result needs to be filtered and monitored long term. If the
result is reliable, the slope could conceivably be used to create an
intermediate controller between the secondary pH controller and the
primary biomass growth rate controller in order to eliminate the change in
the process gain. The controlled variable and manipulated variable of the
intermediate controller would be the slope ($\Delta G/\Delta pH$), and pH,
respectively. The process action would always be reverse for this new
controller because the slope would drop from a high positive value for a
pH below the optimum pH and then drop from zero to a negative value
for a pH above the optimum pH. The result is a triple cascade control

system where the growth rate controller's output is the set point of a slope controller whose output is the set point of a pH controller. The slope controller gain should be scheduled as a function of the slope, with its gain set to zero when the slope is zero. The computed slope would need to be filtered and the intermediate controller response tuned to be slower than the secondary pH controller but faster than the primary growth rate controller.

A bioreactor batch can be characterized by a lag phase, an exponential growth phase, a stationary phase, and a death phase, as shown in figure 4-4e [12]. The plot is an idealized representation of the biomass taken as a whole if the batch cycle was allowed to continue. The phases of individual cells will vary in that some cells may already be dying near the end of the exponential growth phase. Optimization of the batch would have the goal of minimizing the lag phase and maximizing the exponential growth phase. This phase is also known as the logarithmic growth phase since the plot has a nearly constant slope when the vertical axis is the log of the biomass. The batch cycle would end near the start of the stationary phase, and the death phase would be avoided.

In the lag phase, an increase in the number of cells is not noticeable. The length of the lag phase depends on the age, number, and acclimation of the seed cells, and the nutrient content of the fresh medium. The seed cells go through a lag and exponential growth phase in a seed fermenter that can be optimized to harvest cells at the best age in a medium and at conditions similar to the main bioreactor. In order to deal with a different nutrient or higher concentration of the same nutrient, cells take time to produce new or more enzymes, respectively, which increases the duration of the lag phase [12]. The serial use of carbon sources by cells can cause a second lag phase as the preferential carbon source is depleted and the cell diverts its metabolic process from growth to the adjustment of enzymes in order to use the next carbon source [12].

In the exponential growth phase the response of biomass or product concentration is a one-sided integrator because it always increases as it ramps up. Since it is undesirable and impossible to decrease the biomass and product in this phase, it is not possible to control these concentrations directly. However, if the controlled variable is translated from a concentration to a growth or product formation rate, then the one-sided integrating response becomes a two-sided self-regulating response where a steady state can be approached from either direction. The growth rate or product formation rate would be computed as previously described in this section for the purpose of monitoring process sign.

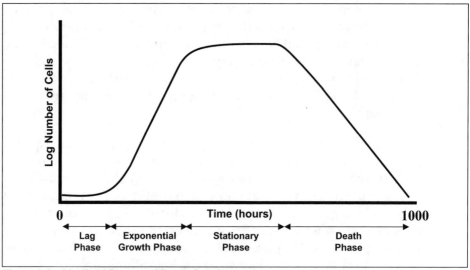

Figure 4-4e. Idealized Phases of Batch Cell Cultivation

Batch profile control requires that the controlled variables be translated from concentration to a rate of change of concentration in order to transform the one-sided integrating response to a two-sided self-regulating response.

Figure 4-4f shows the self-regulating process responses identified for 2x2 MPC where the controlled variables are biomass growth rate and product formation rates and the manipulated variables are substrate concentration and dissolved oxygen. MPC seeks to maintain the growth rate and product formation rate at a set point as shown in figure 4-4g, which in turn can be a function of batch phase time. The penalty on error is increased for the product formation rate relative to the biomass growth rate. With this setup, MPC can provide more consistent and optimized batch profiles for biomass and product concentration, which can reduce batch time and increase yield. Figure 4-4h and 4-4i shows two batches for penicillin production from a simulation running 1000 times faster than real time. In the second batch, the MPC block defined in figures 4-4f and 4-4g was turned on. Section 6-9 in chapter 6 describes the kinetics and operating conditions for this optimization. If a reliable online substrate concentration measurement is not available, MPC can directly manipulate the substrate feed rate. The use of a virtual plant running at 500 times real time for fast identification of these responses and rapid prototyping of MPC applications will be described in chapter 5 [16].

MPC can be set up for simultaneous biomass and product profile control.

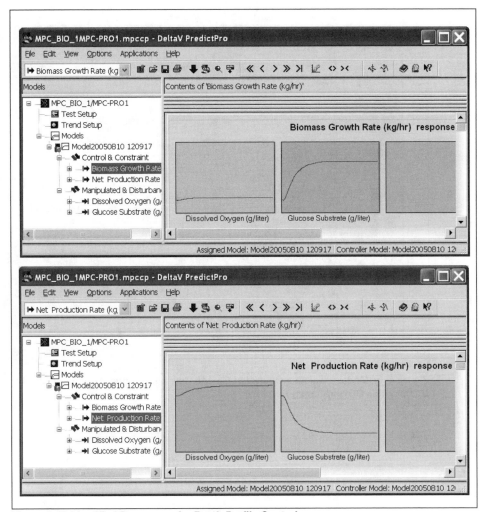

Figure 4-4f. Identified Responses for Batch Profile Control

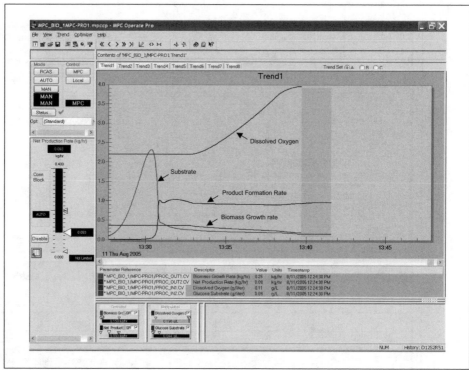

Figure 4-4g. Model Predictive Control of Growth Rate and Product Formation Rate by Manipulating DO and Substrate

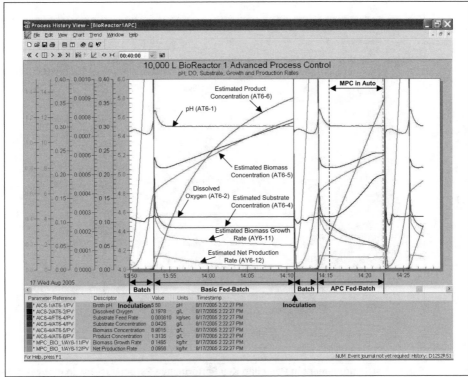

Figure 4-4h. Optimization of Penicillin Batch Profiles by an Innovative MPC

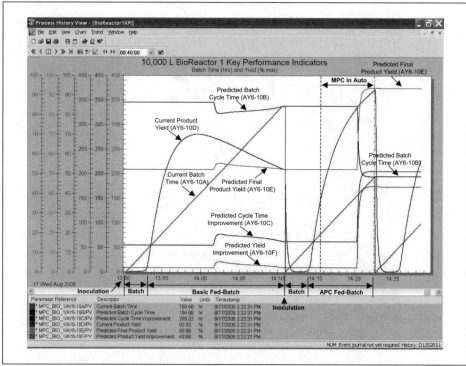

Figure 4-4i. Improvement in Penicillin Key Performance Parameters by an Innovative MPC

References

4-1.   McMillan, Gregory K., and Cameron, Robert, *Models Unleashed: Virtual Plant and Model Predictive Control Applications,* ISA, 2004.

4-2.   Trevathan, Vernon L. (editor), "Chapter 13 - Advanced Process Control." McMillan, Gregory K. (author), *A Guide to the Automation Body of Knowledge.* ISA, 2005.

4-3.   Blevins, Terrence L., McMillan, Gregory K., Wojsznis, Willy K., and Brown, Michael W., *Advanced Control Unleashed: Plant Performance Management for Optimum Benefits,* ISA, 2003.

4-4.   Mehta, Ashish, Wojsznis, Willy, Thiele, Dirk, and Blevins, Terrence L., "Constraints Handling in Multivariable System by Managing MPC Squared Controller." Presentation at ISA EXPO, Houston, October 2003.

4-5.   Alford, Joseph S., "Bioprocess Control: Advances and Challenges." Presentation at CPC-7 to be published in *Computers and Chemical Engineering,* 2007.

4-6.   Callanan, Ken, "A Model Way to Control Fermentation." *Control,* August 2004.

4-7.   Jerden, Cleat, Folger, Tom, and Dee, Matt. "Advanced Control in Small Scale Biotechnology Development." Presentation at Emerson Exchange, Nashville, October 2003.

4-8.   Advanced Application Note. "MPC Implementation Methods for the Optimization of a Slow MV with Good Load Rejection by a Fast MV." Emerson, 2005.

4-9.   Advanced Application Note, "MPC Implementation Methods for the Optimization of the Response of Control Valves for Minimum Variability." Emerson, 2005.

4-10. McMillan, Gregory K., "A Fine Time to Break Away from Old Valve Problems." *Control,* November 2005.

4-11. Yu, Cheng-Ching, and Luyben, William L., "Analysis of Valve-Position Control for Dual Input Processes." *Industrial Engineering Chemical. Fundamentals.* 25, no. 3 (1986).

4-12. Bailey, James E., and Ollis, David F., *Biochemical Engineering Fundamentals.* 2d ed. McGraw-Hill, 1986.

4-13. Nielsen, Jens, Villadsen, John, and Liden, Gunnar, *Bioreaction Engineering Principals.* 2d ed. Kluwer Academic/ Plenum Publishers, 2003.

4-14. Shuler, Michael, and Kargi, Fikret, *Bioprocess Engineering – Basic Concepts.* 2d ed. Prentice-Hall, 2002.

4-15. Alford, Joseph S., Email correspondence with author, 2006.

4-16. Wilson, Grant, McMillan, Gregory K., and Boudreau, Michael, "PAT Benefits from the Modeling and Advanced Control of Bioreactors." Presentation at Emerson Exchange, October 2005.

Chapter 5:
Virtual Plant

Chapter 5

Virtual Plant

5-1. Introduction

The virtual plant is a relatively new concept that is easily confused with existing simulation methods for process design, configuration checkout, and operating training systems. Users may not realize that most of the existing batch process simulations are off line and noninteractive and that most of the real-time dynamic process simulations were originally designed for continuous processes. These real-time process simulations can develop severe numerical errors or even fail under the extreme conditions of batch operations and require interfaces for communicating I/O, controlling inventory, and coordinating with the control system in speeding up, slowing down, pausing, or resuming. The control system engineer is probably most familiar with tieback simulations since these have been predominantly used for configuration checkout and operating training systems. The process response in these tiebacks is mimicked by the trial-and-error adjustment of ramp rates. Ramps are triggered by the manual inclusion of flow path logic to include the opening and closing of valves or the turning on or off of pumps.

Section 5-2 of this chapter discusses the key features that distinguish a virtual plant from existing simulation systems and itemizes the important advantages of the virtual plant approach. Section 5-3 provides a spectrum of uses for virtual plants. Section 5-4 concludes with the basic requirements for and issues concerning implementing a virtual plant. These last two sections note the role of a bench-top system with an industrial control system in getting the most out of a virtual plant. This chapter shows how a virtual plant creates an environment for innovation that is the heart of the Process Analytical Technology (PAT) initiative.

Learning Objectives

A. Appreciate the significance of being able to export and import to the real plant.

B. Recognize the differences between different types of simulations.

C. Know the diverse opportunities that are essential to getting the most from PAT.

D. Realize the importance of a bench-top system with an industrial control system.

E.   Know the basic virtual plant implementation steps.

F.   Determine what process knowledge and measurements are needed.

G.   Be aware of neural network opportunities to supplement first-principle models.

H.   Understand how to use an MPC nonintrusively for adaptation.

I.   Be introduced to the simpler thermodynamic requirements of bioreactors.

J.   Recognize the importance of a charge balance for pH.

5-2.   Key Features

The first key feature that distinguishes a virtual plant (VP) from process simulators is its ability to use the actual configuration, historian, displays, and advanced control tool set of the real plant without translation, emulation, special interfaces, or custom modifications. The configuration database from the real plant can be exported and then imported and downloaded into a personal computer or a control system computer just as if it were an actual hardware controller. Also, files for operator graphics, process history charts, and data history from the real plant can be copied to the computer for the VP so the user has the entire control system of the real plant on a computer, as shown in figure 5-2a [1].

The use of the actual configuration, database, displays, historian, and advanced control tool set without any translation, emulation, special interfaces, or custom modification enables inherent virtual replication of the real control system.

Most dynamic high-fidelity process simulation software offers the ability to build a basic control strategy or sequence inside the simulation environment. However, simulation developers tend to have a process rather than a control background and focus. It is unrealistic to expect the process and batch control capability offered in simulation software to be in the same realm as the control capability of DCS software, which is the culmination of a hundred engineering years (i.e., 5 years by 20 engineers) or more of an effort by process control experts. The overall control functionality in process simulators is primitive compared to the capabilities offered in the modern DCS. The DCS offers capabilities such as sequential function charts and basic function blocks, the batch manager, and advanced control tools like multivariable MPC. These are almost nonexistent in simulators. The effort to duplicate even a simplified version

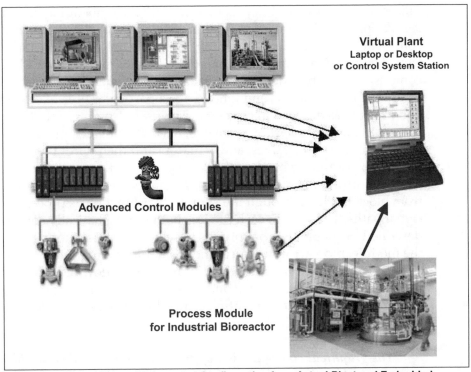

Figure 5-2a. Virtual Plant with Imported Configuration from Actual Plant and Embedded
Process Simulation

of a control system in a dynamic simulation is great. At best, the user ends
up with two control systems with no assurance that they will match well
and no method for automatically managing changes between them.
Consequently, most simulation software now offers a standard or custom
OPC interface. However, in order for the simulations with material
balances to be able to run independently of the DCS for development and
testing, the simulations must still have internal pressure and level loops
set up to prevent volumes from running dry, overflowing, over
pressurizing, or developing flow reversals from pressure gradient
reversals, which can lead to fatal numerical errors [1]. Tables must be
mapped that transfer control from these internal simulation loops to the
DCS loops and initialize the proper controlled variables, set points, and
manipulated variables. DCS loops that do not have a counterpart in the
simulation still need to have their controller outputs initialized. The use of
standard DCS blocks for split-ranged control, velocity limiting, signal
characterization, and signal selection makes the proper initialization of
external simulations problematic.

*The custom programming of control strategies into process
simulations requires significant simplification, effort,
verification, revision, and coordination.*

This leads us to the next key feature in a VP: the embedding of a process simulation in a DCS's configuration modules. For example, the process or control system engineer drags into a module a linked composite block from a simulation library for bioreactors, as shown in figure 5-2b. The composite block has the material balance, component balances, energy balance, charge balance, and kinetics for a particular type of cell culture and product as detailed in chapter 6. The user also drags into the module composite blocks for process streams, mass flowmeters, pumps, fans, transmitters, control valves, and actuators and then wires them up to the bioreactor composite block in the same configuration environment used for the control system. The process information is automatically conveyed from block to block through the "wires" that function as process and signal paths. The flow in each process path is automatically computed, as described in chapter 6, from pressures set for streams entering or exiting the module, bioreactor pressure, and from intervening pump or fan speeds, valve positions, and pressure drops from flow resistance. It is unnecessary to duplicate level and pressure loops for inventory control and sequences for batch operation or to add interface tables for communication of I/O.

The graphical configuration of process modules linked to the control modules eliminates the need to duplicate inventory and batch control or to add interface tables.

Implementing the process simulation as process modules in the configuration database offers advantages beyond the elimination of the previously noted issues of duplicating loops and configuring interface tables. Foremost is the virtual plant's inherent ability to run the process modules at the same real-time execution time multiplier as the control modules. This ensures that the user can set a common real-time multiplier and that the simulation and control system will slow down or speed up in unison. Figure 5-2c shows that the user can set a real-time execution multiplier of 1/30 to 30 for all modules in the VP. When external high-fidelity simulation packages are used, the actual real-time factor may depend on processor loading. This is because of the complexity of the calculations associated with the objectives of process or equipment design. Often during the times of greatest interest, during disturbances or failures, these simulations slow down because the integration step size has decreased for numerical stability and the control system and operator activity have increased. Even if the hook exists between the simulation and DCS real-time multipliers, the DCS is always playing catch-up.

It is difficult to ensure that the control system is running at the same real-time multiplier as a separate process simulation during the times of greatest interest.

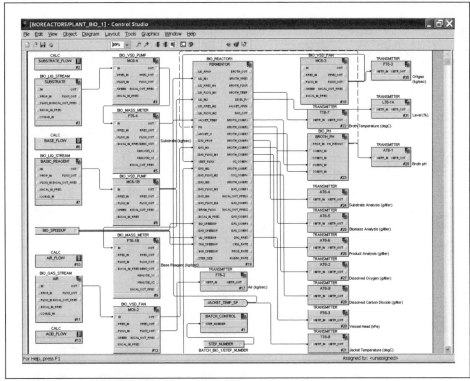

Figure 5-2b. Process Module for Bioreactor

The process modules can be designed to show the proper behavior for failures, startups, shutdowns, and batch sequences. This means that the process modules can handle zero flows (closed valves and turned-off pumps or fans), empty volumes, nonequilibrium conditions, and imperfect mixing. Sophisticated external dynamic process simulations are prone to fatal numerical errors from ill-conditioned matrices used for the simultaneous solution of stream conditions by the pressure-flow solver [2]. These simulations may not only slow down but also shut down during extreme conditions. Consequently, the simulation of batch processes often requires specialized software that runs off line as a single program execution. Functionally, the run conditions are set at the time of execution exactly as was done in the 1970s with a deck of cards. These batch process simulations may appear to the user or be cited by the supplier as interactive, but a change in conditions or parameters requires that the entire program and the batch process be re-executed from start to finish. The ability to include the interaction between the process and the batch and process control strategies is severely limited or impossible. Since these programs are not running real time or a multiple of real time, it is not possible to pause, restore, or play back parts of a batch.

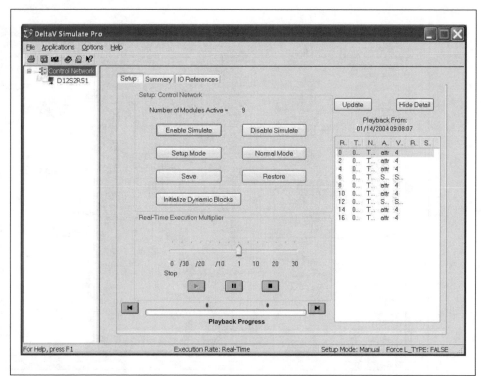

Figure 5-2c. Screen for Control of Virtual Plant Speed, Modes, Capture, and Playback

It is important to verify whether a dynamic simulation can handle the zero flows, empty volumes, and nonequilibrium conditions associated with batch operation.

Most batch process simulation software offered to date is not truly interactive with either the user or the control system.

Process simulations developed by process engineers can predict self-regulating process gains relatively accurately because of the sophistication of the physical properties, thermodynamics, interactions, and equations of state needed for continuous process and equipment design. Process engineers tend to think steady state. Batch process design is noticeably absent in books and courses on chemical engineering. Consequently, these engineers learn and think in terms of a steady state and equilibrium that is consistent with continuous processes.

Most dynamic real-time process simulations were developed by process engineers with steady-state process and equipment design in mind.

Process modules designed by control engineers can be set up to do a better job than process simulations in modeling system dynamics by including the effect of transportation delays, mixing delays, mass transfer rates, kinetics, analyzer cycle times, sensor lags, noise, resolution, and dead band. The result is a closer match to the dead time, time constant, or integrating process gain and a fidelity that is more in tune with the spectrum of uses for control systems [2] [3].

The process simulation dead time is much less than the actual dead time because of the omission of transportation delays, mixing delays, mass transfer rates, kinetics, analyzer cycle time, sensor lags, noise, resolution, and dead band.

The virtual plant also has the ability to simultaneously stop and start the execution and restore and replay simulated conditions for all the control and the simulation modules. Now emerging is the ability to replay actual plant data history files at high speeds in order to adapt and test the process modules without having a connection to the actual plant.

Most simulations used for control system checkout and operator training become obsolete after startup. The investment is lost. The VP offers a better chance of keeping the control system up to date by simply importing the most current configuration and enabling the simulation to better match process changes using the nonintrusive automatic adaptation described in Section 5-4.

Most simulations used for control system checkout and operator training become obsolete after startup because they lack the ability to be automatically updated and adapted.

Finally, in a VP everything is done in the same configuration environment that is used for the actual control system. The focus can be more on the application than learning the inevitable undocumented features and tips and techniques associated with any new simulation software and interface. The advantages offered by a VP are summarized in the following:

1. Control system and graphics do not need to be duplicated, emulated, or translated.

2. Special data interfaces, tables, and initialization issues are avoided.

3. All batch, basic, and advanced control tools can be readily tried out.

4. Controls and simulation can run in unison at the same real-time multiplier.

5. Controls and simulation scenarios can be saved, restored, and played back.

6. Actual plant data can be played back at high speeds for testing and adaptation.

7. Simulations can handle extreme conditions of batch operations and failures.

8. Simulations can incorporate dynamics that are important for tuning and performance.

9. Controls and simulation can stay up to date and have a longer life cycle.

10. Engineers can work in the same environment and focus on the application.

5-3.   Spectrum of Uses

The most familiar use of a virtual plant is for testing and training. For the checkout of batch sequences and the training of operators it is important to be able to repetitively and rapidly simulate batch phases. The ability to stop, start, save, restore, and replay scenarios and record operator actions is critical. For first-pass testing and familiarization of sequences and graphics, an automated tieback simulation may be sufficient. To test and learn about the interaction and performance of both control strategies and the process, the higher fidelity dynamics offered by process modules is important. It opens the door to upgrading the process and control skills of technology, maintenance, and configuration engineers who support operations.

A process simulation with high dynamic fidelity is important for testing process and control system interaction and performance.

Before the configuration even starts in the front end of a project, the process modules can be used to evaluate control strategies and advanced control tools. In the past, this was done with offline dynamic simulations. Having ready access to an industrial tool set for both basic and advanced control and simulations that is adapted to bench-top or pilot-plant runs offers the opportunity for rapid prototyping. This can lead to control definitions that have better detail and potential performance. Bench-top or pilot-plant systems with a mini version of the industrial DCS are now available that greatly facilitate the development and scale-up of the control system [6]. Bench-top systems that have all the functionality of the main manufacturing systems are not yet prevalent because the development groups of these types of companies traditionally do not have the expertise (and more importantly the interest) to configure, maintain, and engineer these systems.

The VP can be demonstrated with the industrial batch, basic, and advanced control systems used in the bench-top or pilot-plant system. The process modules can be adapted by way of a connection to a bench-top or pilot-plant system or by built-in high-speed playback of process data from experimental runs. An opportune time to take advantage of the PAT initiative is during the research and process development phase. The value of bench-top and pilot-plant systems that have an industrial DCS is significant in terms of project schedule, cost, and effectiveness because the DCS allows process control to be designed into the plant at an early stage and puts the biochemist, process technology, and configuration engineer on the same page [6]. However, the best return on investment (ROI) for PAT is realized by eventually implementing advanced process analysis and control in large-scale manufacturing processes.

In a VP, innovative strategies such as effective switching of the controller output (detailed in section 3-5 on "Set-Point Response Optimization" in chapter 3) can be trialed and tuned. Advanced control tools such as adaptive control, auto tuning, fuzzy logic, model predictive control (MPC), neural networks (NN), principal component analysis (PCA), and partial least squares (PLS) can be demonstrated, adjusted, and evaluated "faster than real time".

> *A bench-top or pilot-plant bioreactor that has an industrial control system offers significant opportunities for reducing configuration costs, improving communication, developing a virtual plant, and prototyping advanced control, but the greatest ROI is realized by implementing PAT in large-scale manufacturing processes.*

As a general rule of thumb, five changes are needed in order for each process input to develop an experimental model of a process output. For MPC, this corresponds to a minimum of five step changes in each process input, at least one of which is held long enough for the process output to reach a steady condition. For NN, PCA, and PLS it means a minimum of five batches of process input in which the respective process input differed from the normal value. Although actual plant operation is obviously the best source of data, the long batch cycle time and the desire to minimize disruptions from the introduction of perturbations severely restricts the amount of useful plant data available for the development of MPC, NN, PCA, and PLS. For example, to develop a PCA with four inputs for detecting an abnormal batch would require at least 20 batches with varying inputs and at least five batches with normal inputs. If the batch cycle time is about two weeks, it would take about a year of plant production to have enough data. If you consider that you cannot deliberately optimize the spectrum of variability in the inputs, then it may take several years of production runs to have enough data.

In contrast, perturbations can be automated and introduced to a virtual plant running faster than real time so that in two weeks there is enough data to identify models for MPC, to train NN, to develop latent variables and discriminant analysis for PCA, and to predict economic variables via PLS. The predictive ability of MPC, NN, PCA, and PLS can then be verified and evaluated by the high-speed playback of previous plant batches.

Conventional PCA assumes that all process inputs other than the ones used for the PCA are fixed. A virtual plant running in real time that is synchronized with the actual plant can be used to predict the effect of variations in other process inputs by using the model-based and super model-based PCA algorithms described in section 8-5 of chapter 8.

The virtual plant can also be used to help explore more optimal operating conditions and investigate "what-if scenarios." These scenarios are important for identifying the cause of an abnormal batch. Today, PCA for batch fault detection only identifies a batch as abnormal. Logic needs to be added to diagnose the fault. Fuzzy logic rule sets have been used in conjunction with PCA to provide real-time predictive fault analysis [7]. A VP can help develop these rule sets offline faster than in real time by creating scenarios and evaluating the rule sets through the high-speed playback of previous batches. An online VP synchronized with the real plant can be sped up and run to batch completion in order to predict and analyze abnormal situations based on current batch conditions. A VP can also be used to help create the predictive capability of the PLS "Y space" of economic variables from the PCA "X space" of process variables.

A virtual plant can eliminate process testing and provides years of process data within a few weeks for advanced control, fault analysis, and performance prediction.

Additionally, a virtual plant can be used to provide inferential real-time measurements of important process outputs such as nutrient, biomass, and product concentration. The built-in material balances can be used in conjunction with kinetic models to predict concentrations. These predictions are then delayed so that the values are synchronized with online or lab analysis. The concentration is shifted by a bias that is a fraction of the difference between the inferential and actual measurement, similar to what is done in the feedback correction of a NN used for property estimation [3].

A virtual plant can provide faster, more reliable, and smoother measurements of concentration than can online analyzers.

The rate of change of these concentrations can be used by MPC to optimize batch profiles as described in section 4-4 of chapter 4. A virtual plant can also be run "faster than real time" to batch completion to provide an online prediction of key performance indicators (KPI), such as batch cycle time and yield, based on current batch conditions and inferential measurements.

The virtual plant can speed up the benefits gained from PAT by offering users the ability to use process and endpoint monitoring and control, continuous improvement, and knowledge management tools in an integrated and accelerated manner. The uses of the VP are illustrated in the functional overview provided by figure 1-4b in chapter 1 and are summarized as follows:

A virtual plant can accelerate the benefits of a PAT initiative.

1. Testing configuration and process interactions

2. Educating operators, technicians, and engineers in process and control

3. Rapid prototyping of innovative control strategies and advanced controls

4. Evaluating tuning settings

5. Identifying MPC models

6. Training of NN

7. Developing latent variables and reference trajectories for PCA

8. Developing logic for fault analysis by PCA

9. Predicting abnormal situations on line

10. Making inferential real-time measurements of important concentrations

11. Optimizing batch profiles

12. Predicting batch KPI, such as cycle time and yield, and PLS economic variables

5-4.   Implementation

Probably the biggest obstacle to implementing a virtual plant is the users' lack of knowledge of the kinetic equations that are needed to calculate the growth rates and product formation rates as a function of batch operating conditions, such as temperature, pH, dissolved oxygen, and substrate concentration. The general form of the equations as discussed in chapter 6 can be obtained in some cases from the literature or from internal research. However, bench-top experiments are usually needed to confirm the relationships and to quantify the parameters. Section 5-4 discusses how an MPC can be set up to home in on the values of these parameters if they are in the ball park. If the kinetics is unknown, NN may be able to predict kinetic parameters, terms, or rates from inferential and actual measurements. The growth rates and product formation rates are included in the net calculation of the rate of change of biomass, nutrient, and product mass in the material balance. They are then integrated to get a new accumulation or concentration of the component, as described in chapter 6. Online calculations of oxygen uptake rate (OUR) and carbon dioxide evolution rate (CER) as NN inputs enhance the predictive capability of the NN. The result is a hybrid first-principle and NN model [8]. In general, batch time should not be a NN input because the predictions are undesirably dependent on batch time. A NN without batch time as an input was able to predict ethanol and cell mass [9].

The equations and approximate value parameters for the kinetics as a function of batch conditions are needed to develop and adapt a first-principle model.

The oxygen uptake rates (OUR) and carbon dioxide evolution rates (CER) can be used in lieu of kinetic equations, but it may be difficult to differentiate between the consumption and evolution associated with biomass growth vis-à-vis product formation rates.

If the equations are unknown, a NN may be able to predict the individual growth and production rates if there are online measurements or frequent lab measurements of biomass, product, and substrate concentration [9]. A bench-top or pilot-plant system offers the best opportunity for developing a NN. To minimize the investment, a skid of analyzers could be alternately connected to bench or plant systems to generate the test data needed for NN training and validation. This approach has proved to be successful enough for state governments to accept online NN as a permissible alternative to online analyzers for combustion emissions monitoring (CEM) systems.

Online broth concentration measurements and neural networks can predict growth rates and product formation rates without a prior knowledge of kinetics.

Composite blocks for the bioreactor, streams, final elements, and transmitters are inserted into the process module and wired up. External input and output parameter blocks are added and connected to the final element inputs and transmitter outputs, respectively. The control system configuration is imported, and the analog input blocks in the virtual control system are switched to the simulation mode. The paths of the external inputs for the final elements are chosen via a browser to be the outputs of the analog output blocks of the virtual control system. Similarly, the paths of the external output parameters for the transmitters are selected to be the simulation inputs of the analog input blocks of the virtual control system. Analog outputs and analog inputs of loops that are not modeled can be simulated by simple tiebacks in which the analog output is passed through signal characterizer, dead-time, and filter blocks and then connected to the simulated input of the analog input block.

The concentrations, pressures, and temperatures of each inlet and outlet stream; the flow rate and pressure rise of each pump as a function of speed; the flow coefficient of each valve as a function of position; the agitator pumping rate as a function of speed; and the initial inventory, concentrations, temperature, and pressure of the bioreactor are then set. Neural networks that are developed from bench-top experiments or pilot-plant runs are added as necessary to predict the kinetics described in chapter 6. Control valve dead band and resolution are set in the "actuator" blocks, and measurement delays and lags are set in the "transmitter" blocks.

Bench-top systems offer the best opportunity for developing and validating first-principle or neural network models for kinetics.

The batch sequence, basic process control system, and simulation are run in an offline mode, and the simulation parameters are manually adjusted to match up the virtual with the real plant batch profiles of uncontrolled process outputs and manipulated process inputs. For example, consider how the model may be adjusted for better fidelity for dissolved oxygen control. A kinetic parameter for the oxygen limitation effect is adjusted to match up oxygen uptake rates between the virtual and actual plant. A Henry's Law coefficient is then adjusted that determines the equilibrium dissolved oxygen (driving force for oxygen transfer) to match up pressures. Finally, an oxygen mass transfer coefficient is adjusted to match up air flows.

When representative VP batch profiles are obtained, the processor loading and profile alteration are checked as the real-time execution multiplier is increased, and the maximum practical speed is noted. Key model parameters are chosen as the manipulated variables, and associated process measurements are chosen as the controlled variables of the MPC for adaptation of the virtual plant. An automated test sequence is then run for the MPC at the highest possible speed off line, and the MPC models are identified and visually checked as reasonable in terms of the direction and relative magnitude of the effect [1] [4].

The virtual plant is then connected to the actual plant in a read-only nonintrusive setup. The virtual plant modes, set points, and batch phases come from the actual plant. The set points (targets) of the MPC's controlled variables for adaptation are externally referenced to the key respective measurements of the actual plant, as shown in the overview provided by figure 1-4c in chapter 1 and in the detailed signal paths outlined by figure 5-4a. Since the relationship between model parameters and key process variables is generally nonlinear, a NN can be used to enhance the future trajectories of an MPC. Alternatively, an NN could be trained to adjust model parameters on a one-by-one basis. Since NN are good at interpolation but not extrapolation, it is important to verify whether the inputs are within the training set range of values. The MPC offers a multivariable solution to adaptation of dynamic models [1] [4]. The use of standard tools, such as MPC and NN, are preferred over custom and special reconciliation methods so the focus is on the application rather than the tool development and the results are more maintainable and therefore sustainable. Regardless of which technique is used, the model parameter values must be restricted to be within a practical range.

The model predictive controls and neural networks for adapting the virtual plant can be developed offline and then tested in a read-only mode.

Figure 5-4b shows how an MPC was used to manipulate the Henry coefficient for oxygen to match up a change from 7700 to 7500 kPa per kgmole/m3 at 16:05. This brought the controlled variable of virtual plant head pressure to its set point (target), which is the actual plant head pressure. The speed of adaptation is set by the penalty on move (PM) for the model parameter.

The actual oxygen uptake rate can be measured by temporarily stopping air or oxygen flow and multiplying the average rate of change of the dissolved oxygen (mg/liters per hour) concentration by the broth volume (liters). Alternately, a mass spectrometer can be used to measure the oxygen concentration in the off-gas and a material balance used to measure the net difference of oxygen flow going in and coming out of the broth. If the dissolved oxygen concentration is not constant, the average rate of change of DO multiplied by volume should be included in the calculation.

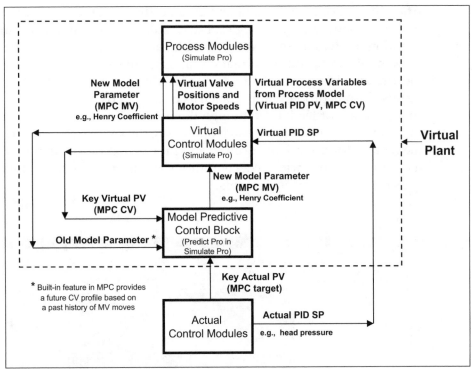

Figure 5-4a. Signal Paths for Nonintrusive Adaptation of Virtual Plant

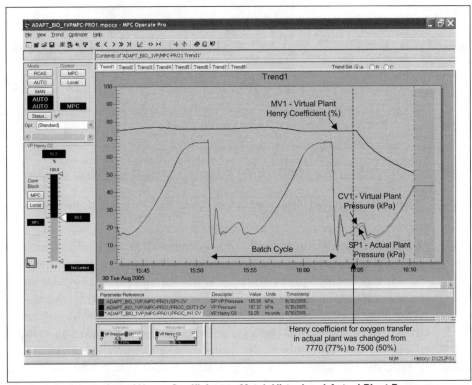

Figure 5-4b. Adaptation of Henry Coefficient to Match Virtual and Actual Plant Pressures

After adaptation, the virtual plant is run offline faster than in real time to prototype new control strategies, PCA, fault detection, and abnormal situation prediction. Inferential variables are computed, such as biomass growth rate and product formation rate, and the automated tests, such as those shown in figure 5-4c, are run to identify the models of these rates versus various process inputs, such as dissolved oxygen and substrate concentration or feed. Even though the process operating point is moving (nonstationary) since it is a batch operation, the software is able to identify the models for optimizing growth and production rates. Inferential measurements of various concentrations are used as needed to fill in the gaps between lab measurements.

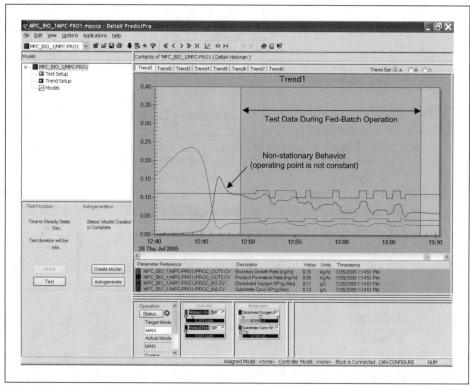

Figure 5-4c. Automated High-Speed Test of Virtual Plant for Model Identification

After innovative control, advanced diagnostics, and MPC applications have been prototyped and documented, they can be implemented on bench-top or pilot-plant systems. If the virtual plant provides inferential measurements, it is run in real time and synchronized with these systems. After verification and validation, and accounting for scale-up factors, these applications and the VP can be installed for plant trials. Initially, these systems run only in the monitoring mode. When process inputs are finally manipulated, they are initially restricted to a very narrow range.

The expertise and time required for using and supporting external high-fidelity process simulation software no longer exists in the process industry, except for some large petroleum and chemical companies. Fortunately, the complexity of these simulations is not needed because for bioreactor modeling it is not of paramount importance that the vapor-liquid equilibrium and physical property data interact to get the thermodynamics exactly right. In fact, relatively simple material and energy balances coupled with a charge balance suffice for a process module. The biggest challenge is getting the kinetics as a function of operating conditions and concentrations to compute the biomass growth, utilization, and production rates.

Excluding kinetics, the thermodynamics needed for models of bioreactors are generally simpler than those needed for chemical reactors.

The charge balance is critical for computing the pH, which is important for the kinetics. Process simulations in the literature for bioreactors generally use empirical relationships for pH that do not show the effect of alternative operating conditions and upsets. Chapter 6 shows how to set up the charge balance from the component balances.

A charge balance is needed to properly simulate pH.

References

5-1.   McMillan, Gregory K., and Cameron, Robert, *Models Unleashed: Virtual Plant and Model Predictive Control Applications.* ISA, 2004.

5-2.   Boyes, Walt, Hebert, Dan, O'Brien, Larry, and McMillan, Gregory K., "The Light at the End of the Tunnel Is a Train (Virtual Plant Reality)." *Control*, August 2005.

5-3.   Trevathan, Vernon L. (editor), "Chapter 12 - Process Modeling." McMillan, Gregory K. (author), in *A Guide to the Automation Body of Knowledge.* ISA, 2005.

5-4.   Blevins, Terrence L., McMillan, Gregory K., Wojsznis, Willy K., and Brown, Michael W., *Advanced Control Unleashed: Plant Performance Management for Optimum Benefits.* ISA, 2003.

5-5.   Mansy, Michael, McMillan, Gregory K., and Sowell, Mark, "Step into the Virtual Plant." *Chemical Engineering Progress*, February 2002: 56-61.

5-6.   BioNet Bioreactor Control System. Broadley-James Bulletin L0016B, 2003.

5-7.   Fickelscherer, Richard J., Lenz, Douglas H., and Chester, Daniel L., "Fuzzy Logic Clarifies Operations." *Intech*, October 2005.

5-8.   Silva, R. G., Cruz, A. J. G., Hokka, C. O., Giordano, R. L. C., and Giordano, R. C., "A Hybrid Feedforward and Neural Network Model for the Cephalosporin C Production Process." *Brazilian. Journal of Chemical Engineering.* 7, no.4-7, December 2000.

5-9.   Pramanik, K., "Use of Artificial Neural Networks for Prediction of Cell Mass and Ethanol Concentration in Batch Fermentation Using Saccharomyces cerevisiae Yeast." *IE (I) Journal – CH* vol. 85, September 2004.

Chapter 6:
First-Principle Models

Chapter 6

First-Principle Models

6-1. Introduction

Recently, a fed-batch fermentation simulation was developed by the Department of Chemical and Environmental Engineering at the Illinois Institute of Technology and made available to the public as a test bed for control and optimization studies. This model is the basis of a virtual penicillin production plant developed to run on a distributed control system (DCS). Elements of this test-bed model were incorporated into a fed-batch model of human epidermal growth factor (hEGF) production by recombinant bacteria; and, into a batch model of simultaneous saccharification and fermentation (SSF) by baker's yeast.

In this chapter, a model of a cell culture that divides the cell mass into four units or pools is presented to demonstrate the increase in the complexity of mass balances when a model considers cell structure. Section 6-2 begins our discussion of virtual bioreactors by pinpointing our location on a map of model types. Section 6-3 shows how mass, energy, and component balances are used to create a virtual plant. Section 6-4 gives the heat of reaction used in energy balances in all three simulations. Section 6-5 shows how pH can be calculated in a fermentation broth by a charge balance on weak acids and weak bases. Section 6-6 provides the values of the constants and initial values of the variables used in all three virtual plants. Section 6-7 on kinetics presents the calculations of the specific growth rate of each culture and shows how product formation is correlated to specific growth rate. Section 6-8 discusses mass transfer issues in microbial cultures and mammalian cell cultures. On-line methods for calculating the volumetric oxygen transfer coefficient are also presented. Section 6-9 shows how all these equations come together in batch profiles of the penicillin, SSF, and hEGF simulations.

Learning Objectives

A. Understand mass and energy balances in dynamic, sequential modular simulations.

B. Calculate pH in a bioreactor from a charge balance.

C. Learn how to create detailed unstructured and structured bioreactor models.

D. Learn how to estimate maximum specific growth rate and volumetric mass transfer coefficients.

E.   Understand gas-liquid mass transfer issues in different bioreactor types.

6-2.   Our Location on the Model Landscape

Most of the dry mass of a cell is protein. Proteins can self-assemble into large structures and can combine to form machines like molecular pumps. Enzymes simultaneously catalyze acid and base reactions. Enzymes can form complexes that increase the rate and efficiency of metabolic reactions. Signal and regulatory proteins coordinate cellular activity and add regulatory layers to gene expression [1]. Proteins are in turn regulated by posttranslational modifications that depend on the state of the cell [2].

Only a subset of a genome and cellular protein interactions can be modeled in whole cell simulations. One simulator that can do this, E-CELL, simulates a subset of these interactions in *Mycoplasma genitalium*, an organism that has one of the smallest known chromosomes [3]. High-fidelity models of a cell enable researchers to conduct experiments *in silico* that improve researchers understanding of cellular systems and enhancing drug discovery.

To simulate the relationship between substrate utilization, cell growth, and product formation in a bioreactor, it is not necessary to use complex and expensive theoretical models of cellular structure and function like E-CELL. Bioreactor models that are used for design, control, and process optimization or for training depend instead on parameters that are fit to experimental data. Two common experimental parameters in bioreactor simulation are the oxygen mass transfer coefficient, $K_l a$, and the maximum specific growth rate of a cell, μ_{max}. Techniques to determine their values for a given bioreactor configuration and a particular microbe or cell type are discussed in section 6-8 of this chapter. Seborg et al. [4] refers to this type of model as "semi-empirical." Semi-empirical models extrapolate better than purely empirical models, but do not have the high cost of a whole cell model.

Bioreactor models are further delineated by the amount of structure and segregation [5]. A model that considers the biomass as a single variable is unstructured. The yeast and fungus fermentation simulations presented in this chapter are unstructured. Since all cells are considered to be identical, these models are also unsegregated.

A third simulation describes the growth and product formation of a recombinant bacterium. As foreign protein production is induced, the cells are segregated based on their plasmid content and their ability to reproduce.

A structured kinetic model used to describe hybridoma growth and monoclonal antibody production divides the cell mass into amino acid, nucleotide, protein and lipid pools.

In this book, cellular structure and function are not modeled on first principles as known to biologists. First-principle models, as used in this book, refer to chemical engineering first-principle models using material, energy, and charge balances in sequential modular simulations. Simulation elements such as pumps, valves, and reactors are connected by streams that are broken into components based on mass fraction. The mass and energy balances on each element in a stream are solved in sequence.

6-3.   Mass, Energy, and Component Balances

Four case studies of industrial scale bioreactors are presented in this chapter. Penicillin production is based on the unstructured model of Birol et al. [6]. Ethanol production by simultaneous saccharification and fermentation of corn starch by yeast is based on a Purdue model [7].

Human epidermal growth factor (hEGF) production on a recombinant *Escherichia coli* is based on a kinetic model by Zheng et al. [8]. A bioreactor simulation of monoclonal antibody production is proposed. The ethanol bioreactor runs in batch mode. The other three bioreactors operate in fed-batch mode. In batch mode, all the media is charged to the bioreactor before it is inoculated. In fed-batch mode, a substrate is fed continuously to the bioreactor after some initial batch period.

The entire bioreactor flow sheet is represented by a sequential modular simulation. Processing elements are represented by standard modules and bioreactor modules are customized for each culture type. The simulation runs on a DCS and each module is a custom block. Because the simulations run on a DCS platform, state-of-the-art control strategies can be used in a straightforward manner to configure and interface to these virtual plants.

Sequential modular simulation, which originated in the 1950s, is the basis of many commercial simulators [9]. Because physical property data is used to solve mass and energy balances, it is referred to as a first principle simulation.

Each processing element in a bioreactor system—variable-speed pumps, meters, valves, and vessels—is represented by a module. Modules are connected as in a flow sheet, and each module is solved in sequence. Each element in a process stream affects the downstream operating conditions. For instance, if a valve closes, the stream flow stops.

The properties needed in mass and energy balances for a given processing element are defined in parameters within each module type. To compute mass and energy balances, all modules define the composition of the processing stream with one or more component. The composition of a multicomponent stream is based on mass fractions, not volume fractions.

Table 6-3a. Physical Properties Data for a Single Module

Component Name	Molecular Weight	Density	Specific Heat	Boiling Point	Heat of Vaporization	Heat of Reaction	Reaction Coefficient	Activation Energy
Component 1								
Component 2								
Component 3								
Component 4								
o								
o								
Component Z								

Table 6-3b. Mass Fraction of Stream Components

Possible Components	Mass Fraction
Component 1	0
Component 2	0
Component 3	0.2
Component 4	0.2
o	
o	
Component N	0.6

Modules are categorized as actuating or processing depending on their action on stream components. Actuating modules have no composition change. Module input and output contain the same components and mass fractions. In processing modules, components can be mixed with no reaction as in a vessel. Also, components can be transferred between streams, or components can be reacted with a net change in mass fractions within a single processing stream. A bioreactor is a single processing module that has multiple input and output process streams.

The following is a list of actuator modules used in bioreactor simulations:

GAS_STREAM
LIQUID_STREAM
MASS_METER

VSD_FAN
VSD_PUMP
TRANSMITTER.

Processing modules include the various bioreactors as well as a module to regulate charge balance:

PH
PENICILLIN
YEAST
rBACTERIA
HYBRIDOMA.

Figure 6-3a shows the internals of a type of actuator module: a variable-speed drive fan. The module is designed to be inserted in a gas stream that contains oxygen, carbon dioxide, water vapor, nitrogen, and helium. This module calculates an energy balance that increases the outlet pressure. Components and their mass fractions do not change.

A bioreactor module represented by a composite block is shown in figure 6-3b. It contains two liquid streams, four gas-feed streams, and bioreactor off-gas and broth outlet streams. Since the simulations presented in this text are batch or fed-batch, the broth flow is zero. Liquid streams may include substrate feeds, and gas feed streams may include oxygen, carbon dioxide, ammonia, inert gas, and air.

If the entire batch process cycle was simulated, the broth flow could be connected to a harvest process stream, and steam-in-place (SIP) and clean-in-place (CIP) streams could be added to the bioreactor module.

All bioreactor modules contain transfer calculations similar to those in the VSD_Fan actuator module in figure 6-3a. The bioreactor modules contain additional calculations for more extensive mass and energy balances and interphase mass transfer of oxygen. The structured or unstructured kinetic models described earlier also reside in calculations within the module.

In sequential modular simulations, actuating and processing modules are connected just as they are shown on a process flow sheet. For instance, a substrate line to a bioreactor may include pump, control valve, flow meter, and block valve modules with a vessel or a specified pressure as the source of the stream and a bioreactor at the stream termination. These modules can be solved in a sequence and converge to a solution in one iteration if there is no recycle.

In the case of a recycle stream, a material balance on Module 1 in figure 6-3d cannot be made until the recycle stream is known. However, Module 3 must be computed to get values in the recycle stream. Module 3 depends

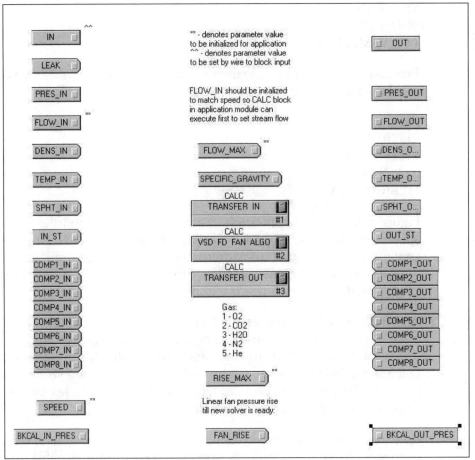

Figure 6-3a. Internals of a Module

on Module 2, which in turn depends on Module 1. To resolve the loops of information created by the recycle stream, a convergence block decouples the interconnections and allows information to flow sequentially.

Decoupling recycle interactions is called *tearing*. It is an iterative process in which the mass and energy values for all modules are calculated from an initial guess. A new guess is based on the result of this calculation, and the process is repeated until the difference between guesses is within a predefined tolerance.

In this dynamic simulation system, the tear concept is used to establish flows and pressures into and out of processing vessels. Conditions in the processing vessels are assumed to be constant, while multiple iterations of flow calculations are performed in process stream modules. A convergence block is placed between the last module in the process stream and the downstream vessel. It calculates the stream flow guess at each

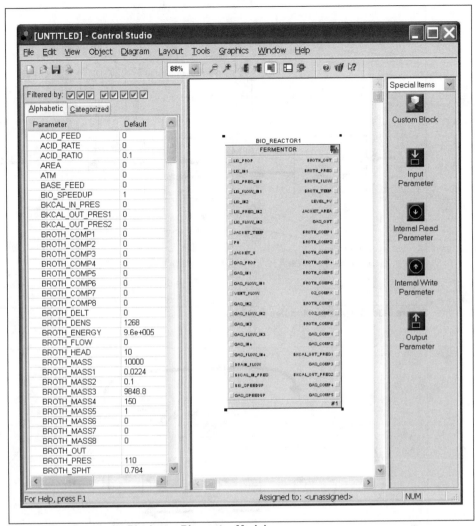

Figure 6-3b. Composite Block as a Bioreactor Module

iteration. The stream flow is established by this iterative process between each execution of a vessel simulation.

Closing a block valve in a stream simulation requires special consideration. It is handled by propagating the downstream pressure counter to the stream flow in all the modules between the block valve and the downstream vessel. Figure 6-3a shows the back-calculation inlet and outlet pressure parameters for the VSD_Fan module.

The flow sheet in figure 6-3e is representative of the bioreactor that produces penicillin. In the sequential modular simulation of this bioreactor, each component in the flow sheet is represented by a module.

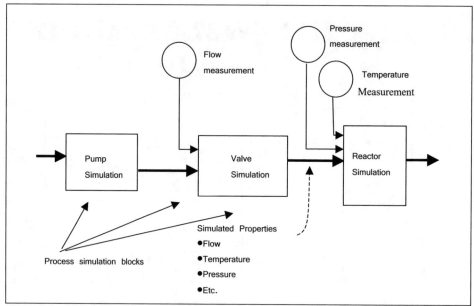

Figure 6-3c. Simulation Modules in Series

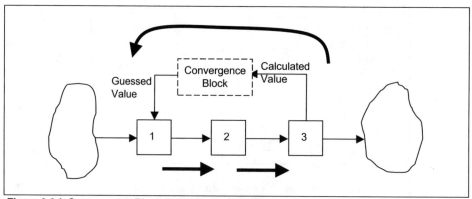

Figure 6-3d. Convergence Block to Decouple Interactions

6-4.   Heat of Reaction

An energy balance on the bioreactor is an integral part of the sequential modular simulation. To complete the system energy balance, the heat generated by microbial reactions must be estimated. Birol et al. [6] provide a volumetric heat production rate with separate biomass growth and maintenance terms.

$$\frac{dQ_{rxn}}{dt} = Y_{q1}\frac{dX}{dt}V + r_{q2}XV \qquad (6\text{-}4a)$$

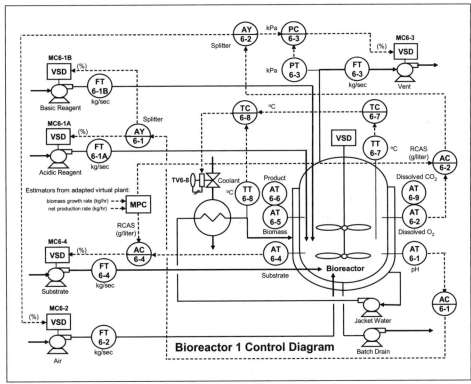

Figure 6-3e. Flow Sheet of a Penicillin Production Bioreactor

Birol et al. [6] obtain the yield of heat generation on biomass growth, Y_{q1}, from Nielsen et al. [5] and estimate a heat generation constant for maintenance based on the evolution rate.

6-5. Charge Balance

The dynamic simulation contains a separate module for pH. It is similar to a metering module in that it does not modify the process mass and energy balances. The only output is the vessel or stream pH. In the simulations in this book, it can be thought of as a sensor connected to the bioreactor module.

The pH of a culture medium is calculated by balancing the charges of all ions in solution. The net charge must always be zero. In all the simulations presented in this book, both the metabolic by-products of cells that drive the pH from an optimum and the acids or bases that neutralize the metabolic by-products are weak. Not all ions are in solution. Relationships between ion concentration and dissociation constants given in McMillan, et al. [10] are used by this pH module to calculate the ions in solution. The module can accommodate up to three dissociations per molecule, but only single dissociations occur here. The equations require acid and base concentration in normality units, and they are solved only in terms of logarithmic acid dissociation constants (pKa).

$$N_1 = \frac{1}{1+P_1} sN \qquad\qquad (6\text{-}5a)$$

where:

$$P_1 = 10^{(s*(pH-pK_1))} \qquad\qquad (6\text{-}5b)$$

For weak acids and bases that have a single dissociation, the ion concentrations depend on pH and are computed with the following charge balance:

$$\sum_i \left[\frac{1}{1+P_{bi}} N_{bi}\right] - \sum_i \left[\frac{1}{1+P_{ai}} N_{ai}\right] + 10^{-pH} - 10^{(pH-pKw)} = 0 \qquad (6\text{-}5c)$$

where:

N	=	concentration of an acid or base (normality)
N_1	=	concentration of ions from a single dissociation (normality)
N_{ai}	=	concentration of an acid i (normality)
N_{bi}	=	concentration of a base i (normality)
pK_1	=	negative base 10 logarithmic first acid dissociation constant
pK_w	=	negative base 10 logarithmic water dissociation constant
P_1	=	parameter for an acid or base with one dissociation
P_{ai}	=	parameter for dissociation of an acid i with a single dissociation
P_{bi}	=	parameter for dissociation of a base i with a single dissociation
s	=	ion sign (s = -1 for acids and s = +1 for bases)

An interval-halving algorithm for finding the pH that solves the charge balance for all ionic species is listed in the appendix to the ISA book *Advanced pH Measurement and Control*. The pH module in the sequential modular simulation uses this charge balance. The charge balance computation is simple and well suited to dynamic simulation and it can resolve pH over a large concentration range.

Microbes and cell cultures are very sensitive to pH. Proteins and especially enzymes depend on a three-dimensional or tertiary structure in order to function. This structure is formed by hydrogen and ionic bonds that are broken by extremes of hydrogen and/or hydroxyl ion concentration. Therefore, all cells must maintain a constant pH within the cytoplasm to maintain enzyme activity.

Cells can correct for small variations in environmental pH by pumping potassium or sodium ions across the cell membrane. Gibbs free energy is needed to maintain the proton gradient across the membrane [5]. As the environmental pH moves further from the growth optimum, *Escherichia coli* synthesizes chaperons or acid shock proteins to prevent acid denaturing of protein and to refold denatured proteins [6-11]. The increased cellular resources needed to maintain pH homeostasis within the cytoplasm cause the cell growth and product formation rate to fall.

The optimum culture media pH for ethanol production by *Saccharomyces cerevisiae* and for penicillin production by *Penicillium crysogenum* is about 5.0 [6] [11]. Birol et al [6] assume direct hydrogen ion transport across the cell membrane. The amount of hydrogen ion excreted is related to the biomass growth according to the expression given in Birol et al. [6].

$$\frac{d[H^+]}{dt} = \gamma\left(\mu X - \frac{FX}{V}\right) + \left[\frac{-B + \sqrt{(B^2 + 4x10^{-14})}}{2} - [H^+]\right]\frac{1}{\Delta t} \quad \text{(6-5d)}$$

where B is defined as:

$$B = \frac{\left(\frac{10^{-14}}{[H^+]} - [H^+]\right)V - C_{a/b}(F_a + F_b)\Delta t}{V + (F_a + F_b)\Delta t} \quad \text{(6-5e)}$$

In both simulations, pH is maintained at its optimum by adding ammonia. This empirical equation was determined by measuring the amount of ammonia needed to maintain pH in a culture as the biomass increases. This hydrogen ion concentration is used directly in a growth rate inhibition term.

The *Saccharomyces cerevisiae* simulation describes the simultaneous saccharification and fermentation of cellulose and of starch to fuel ethanol. Enzyme hydrolysis and fermentation in the same vessel at the same time is less expensive than separate reactions and helps to overcome glucose inhibition of hydrolysis. Mosier and Ladisch [7] point out the necessity of matching the pH optima of the hydrolysis and the fermentation. Manufacturers' literature gives α-amylase a pH optimum of 5.5 and amyloglucosidase a pH optimum near 4.5. A starch-to-ethanol SFF can be expected to run at about pH 5.0.

The simulation of human epidermal growth factor (hEGF) expression on recombinant *Escherichia coli* is based on a kinetic model proposed by Zheng et al. [8]. This model does not consider the inhibitory effect of acetic acid production on *E. coli* growth. The differential equation describing acetic acid formation by aerobically grown *E. coli* on glucose in a fed-batch culture proposed by Roeva et al. [12] provides the input to the charge balance and models the growth inhibition.

$$\frac{dAc}{dt} = \frac{1}{Y_{Ac/X}} \mu X - \frac{F}{V} Ac \qquad (6\text{-}5f)$$

Again, ammonia in aqueous solution is used to neutralize the acid produced as metabolic by-product. Since the hydrogen ion concentration is not calculated directly, as in the ethanol and penicillin fermentations, acetate and ammonia concentrations in the culture media are input to the weak acid–weak base calculation in the pH module.

In a mammalian cell culture there are two principle substrates: glucose and the amino acid glutamine. Unlike bacteria, yeast, and fungi, mammalian cells also require amino acids in their culture medium. Lactic acid is a by-product of glucose and glutamine metabolism and ammonia is a by-product of glutamine metabolism [13].

Animal cells cannot accommodate excursions in the environmental pH to the same extent as bacteria and fungi. The culture media must be maintained in a narrow pH range, typically 6.9 to 7.3. Sodium bicarbonate is added to buffer the media. In the early stages of the culture, carbon dioxide is added to the bioreactor gas mixture. An equilibrium relationship based on Henry's law establishes the concentration of dissolved carbon dioxide. The dissolved carbon dioxide establishes equilibrium with the bicarbonate and acts as an acid in the buffering system.

As cells grow, it is primarily the lactic acid formation that causes pH to decrease. The bioreactor gas stream is used to strip carbon dioxide off the culture medium. pH control is maintained by decreasing dissolved carbon dioxide.

6-6.   Parameters and Their Engineering Units

All bioreactors presented in this section are simulated at an industrial scale. That means the working volume is about 10,000 liters or 10 cubic meters.

Table 6-6. Typical Units used in Simulations

Physical Property	Units
Mass	kilogram
Volume	Cubic meter
Concentration	Kilograms/cubic meter
Concentration in Henry Equation for dissolved oxygen and dissolved carbon dioxide	Kg-mole/cubic meter
Time	second
Reaction or mass transfer rates	Kilograms/second
Specific growth or specific production rates	/second
Energy	Kilojoules
Temperature	Degrees Celsius

6-6.1.   Parameters Describing the Model of Penicillin Production in Fed-Batch Fermentation

Table 6-7a. Penicillin Model Variables [6]

Symbol	Units	Description
C_L	Kg/m^3	Dissolved oxygen concentration
CO_2	Kg/m^3	Carbon dioxide concentration
F	Kg/s	Feed flow rate
F_a	Kg/s	Acid Feed flow rate
F_b	Kg/s	Base Feed flow rate
$[H^+]$	Kg/m^3	Hydrogen ion concentration
$k_l a$	s^{-1}	Volumetric oxygen transfer coefficient
N_i		Number of impellers
P	Kg/m^3	Penicillin concentration
P_G/V_L		Power input per unit volume of bioreactor
Q_{rxn}	KJ	Heat of reaction in bioreactor
S	Kg/m^3	Substrate concentration in the bioreactor
S_f	Kg/m^3	Feed substrate concentration
T	$°C$	Bioreactor broth temperature
T_f	$°C$	Feed temperature of substrate
V	m^3	Culture volume
V_s	m/s	Superficial air velocity
X	Kg/m^3	Biomass concentration in the bioreactor
μ_{pp}	s^{-1}	Specific rate of penicillin production

Table 6-7b. Penicillin Model Constants [6]

Symbol	Units	Description	Value
b		Constant	0.60
$C_{a/b}$	$Molar$	Concentration of acid and base – assumed to be equal	3
E_d	$Kcal/Kg$	Activation energy for cell death	209200
E_g	$Kcal/Kg$	Activation energy for cell growth	21338
K	s^{-1}	Penicillin hydrolysis rate	1.11×10^{-5}
k_g		Arrhenius constant for growth	$7.0 \times 10^{+3}$
k_d		Arrhenius constant for death	10^{+33}
K_I	Kg/m^3	Inhibition constant for product formation	0.10
K_p	Kg/m^3	Product inhibition constant	0.0002
K_X	$mole/L$	Contois saturation constant	0.15
K_1	$mole/L$	Constant in hydrogen ion inhibition	10^{-10}
K_2	$mole/L$	Constant in hydrogen ion inhibition	7.0×10^{-5}
K_{op}		Product limitation on oxygen constant	0.0005
K_{ox}		Biomass limitation on oxygen constant	0.02
m_0	s^{-1}	Maintenance coefficient on oxygen	1.30×10^{-4}
m_x	s^{-1}	Maintenance coefficient on substrate	3.89×10^{-6}
P		Exponent on the dissolved oxygen product inhibition term	3
r_{q2}	$KJ/Kg/s$	Heat generation for maintenance	1.95×10^{-7}
$Y_{P/O}$	Kg/Kg	Penicillin / Oxygen yield constant	0.2
$Y_{P/S}$	Kg/Kg	Penicillin / Glucose yield constant	0.9
Y_{q1}	KJ/Kg	Yield of heat generation on biomass growth	251
$Y_{X/O}$	Kg/Kg	Biomass / Oxygen yield constant	0.04
$Y_{X/S}$	Kg/Kg	Biomass / Glucose yield constant	0.45
α_1	$mole/Kg$	Constant relating carbon dioxide to growth, moles CO_2 per Kg biomass.	0.143
α_2	$mole/Kg/s$	Constant relating carbon dioxide to maintenance energy, moles CO_2 per Kg biomass per second.	1.11×10^{-10}
α_2	$mole/m^3/s$	Constant relating carbon dioxide to penicillin production, moles CO_2 per m^3 of broth volume per second.	2.78×10^{-8}

x		Constant in correlation between oxygen transfer and power per unit volume	2.0
σ		Constant in correlation between oxygen transfer and power per unit volume	2.8
y		Proportionality constant: $[H^+]/g$ biomass	10^{-5}
λ		Constant in evaporative heat loss calculation	6.94×10^{-8}
μ_x	s^{-1}	Maximum specific growth rate	2.56×10^{-5}
μ_p	s^{-1}	Maximum specific rate of penicillin production	1.39×10^{-6}

6-6.2. Parameters Describing the Model of hEGF Expression in Fed-Batch Cultures of Recombinant *Escherichia coli*

Only parameters that do not already appear in the penicillin fed-batch model are listed in this section.

Table 6-7c. hEGF Model Variables [8]

Symbol	Units	Description
Ac	Kg/m^3	Acetate concentration
P_{ex}	Kg/m^3	Concentration of extracellular proteins
P_f	Kg/m^3	Foreign protein concentration
X	Kg/m^3	Total biomass
X_+	Kg/m^3	Dividing and plasmid-bearing cells
X_-	Kg/m^3	Dividing and plasmid-free cells
X_n	Kg/m^3	Viable-but-non-culturable cells
X_l	Kg/m^3	Lysed cells
I_e	Kg/m^3	Extracellular inducer concentration
I_i	Kg/m^3	Intracellular inducer concentration
$k_L^{O_2}a$	s^{-1}	Volumetric oxygen transfer coefficient
α	s^{-1}	Conversion rate of the X+ to Xn
μ_+	s^{-1}	Specific growth rate of X+
μ_-	s^{-1}	Specific growth rate of X_
π	s^{-1}	Specific foreign protein production rate

Table 6-7d. hEGF Model Constants [8]

Symbol	Units	Description	Value
A		Constant	0.000255
B	Kg/m^3	Constant	0.203
C	s^{-1}	Constant (h_1)	2.194×10^{-7}
D		Constant	0.95
E	Kg/m^3	Constant	0.88
K_π	Kg/m^3	Inducer saturation constant for product synthesis	0.028
k_1		Constant	0.53
k_2	Kg/m^3	Constant	0.162
k_3		Growth associated product yield	1.37
k_4	s^{-1}	Non-growth associated product yield	5.17×10^{-8}
k_5	s^{-1}	Foreign protein degradation rate	1.39×10^{-7}
$k_L^{CO_2} a$	s^{-1}	Volumetric CO_2 transfer coefficient	0.02217
m	s^{-1}	Maintenance coefficient	6.67×10^{-6}
$Y_{Ac/X}$	Kg/Kg	Yield coefficient for acetate	0.150
$Y_{CO_2/X}$	Kg/Kg	Carbon dioxide yield on biomass	0.512
$Y_{DO/X}$	Kg/Kg	Dissolved oxygen yield on biomass	0.683
β	s^{-1}	Lysis rate of X_	3.06×10^{-7}
γ	s^{-1}	Lysis rate of X+	4.17×10^{-7}
δ	s^{-1}	Lysis rate of Xn	5.83×10^{-6}
θ		Probability of forming a plasmid-free cell	0.01
μ_{m+}	s^{-1}	Maximum specific growth rate of X+	0.000137
μ_{m-}	s^{-1}	Maximum specific growth rate of X_	0.000136
φ		Conversion factor of lysed biomass to extracellular proteins	0.5

6-6.3. Parameters Unique to Simultaneous Saccharification and Fermentation Model

Only parameters that do not already appear in the penicillin fed-batch model are listed in this section.

Table 6-7e. Ethanol Model Variables [7]

Symbol	Units	Description
C	Kg/m^3	Concentration of cellulose in the bioreactor
v	s^{-1}	Specific ethanol production rate

Table 6-7f. Ethanol Model Constants [7]

Symbol	Units	Description	Value
E	Kg/Kg	Efficiency of substrate utilization for cell mass production	0.249
K_{i1}	mM/L	Michaelis-Menten inhibition constant, cellulose hydrolysis	0.50
K_{i2}	mM/L	Michaelis-Menten inhibition constant, cellobiose hydrolysis	1.22
K_{m1}	mM/L	Michaelis-Menten constant, cellulose hydrolysis	0.08
K_{m2}	mM/L	Michaelis-Menten constant, cellobiose hydrolysis	2.50
K_s	Kg/m^3	Monod Constant	0.315
n		Toxic power constant in ethanol inhibition	0.36
P_m	Kg/m^3	Maximum product concentration	87.5
V_{m1}	mM/s	Maximum conversion rate cellulose to cellobiose	0.34
V_{m2}	mM/s	Maximum conversion rate cellobiose to glucose	0.58
v_m	s^{-1}	Maximum specific production rate	3.19×10^{-4}
$Y_{P/S}$	Kg/Kg	Yield of ethanol on glucose	0.434
$Y_{X/S}$	Kg/Kg	Yield of yeast on glucose	0.07

6-7.   Kinetics

Many types of bioreactors exist: packed bed, fluidized bed, airlift, immobilized cell, and plug flow and cell culture wave bioreactors. The activated sludge process is an example of a continuous bioreactor with a biomass recycle. At the production scale, stirred tanks are still the most common bioreactor type.

Stirred tanks can be operated in batch, fed-batch, and continuous modes. Nielsen et al. [5] list the advantages and disadvantages of each type. Continuous stirred tank reactors are powerful laboratory tools for biological investigations into the biochemistry, ecology, and physiology of microbes. Their susceptibility to infection and mutation makes them poor candidates for production-scale operation. Their few industrial applications are lactic acid and single cell protein production.

Batch stirred tank reactors reduce the risk of contamination and the growth limiting substrate is utilized completely. This is an advantage in the conversion of cellulose or corn starch to fuel ethanol.

Sometimes the metabolic by-product of interest or the biomass growth itself is repressed under the presence of too much substrate. This leaves the batch bioreactor at a disadvantage since all the substrate must be charged to the bioreactor before it is inoculated. The fed-batch mode of operation avoids this catabolic repression by feeding the substrate to the bioreactor over the course of the production run. The bioreactor is inoculated with all the nutrients needed for the production run but with a smaller amount of the growth-limiting substrate. After inoculation, the bioreactor is run in batch mode during the lag and initial growth phases. Additional substrate is charged to the bioreactor either intermittently or continuously. An intermittent feed usually refers to periodic batch charges of substrate during the fed-batch portion of the production run. A fed-batch bioreactor can be set to a quasi-steady state if the substrate is fed to the bioreactor continuously at the same rate at which it is consumed by the cells.

Penicillin Production
The simulation of penicillin production presented in this book is based on an unstructured, unsegregated model offered by Birol et al. [6] as a test-bed, fed-batch fermentation process. The model includes the effects of environmental variables on biomass growth and product formation and is well suited to optimization studies.

When a bioreactor is in fed-batch mode, a general mass balance on the biomass is represented by equation 6.7a.

$$V\frac{dX_{biomass}}{dt} = \frac{dV}{dt}(X_{feed} - X_{bioreactor}) + \mu X_{bioreactor} \quad (6\text{-}7a)$$

Generally, there is no biomass in the feed line, and the mass balance can be written in terms of only the bioreactor biomass.

$$V\frac{dX}{dt} = -\frac{dV}{dt}X + \mu X \quad (6\text{-}7b)$$

All that is needed to solve this expression are the initial conditions and a value for the specific growth rate, μ. Biomass itself, hydrogen ion concentration, dissolved oxygen, temperature, and substrate can all affect the specific growth rate. It is the key variable in the equation.

The Monod kinetics, described in Aiba et al. [14], Bailey and Ollis [15], Nielsen et al. [5], and Shuler et al. [16], is one of the earliest and most often used equations for describing the relationship between substrate concentration and biomass concentration. The autocatalytic nature of cell growth gives a sigmoid growth curve in a batch bioreactor. Monod kinetics generally gives a good fit to the growth curve.

$$\mu = \frac{\mu_x S}{K_x + S} \tag{6-7c}$$

The two new constants are the maximum specific growth rate, μ_x, and the saturation constant, K_x. If the concentration of substrate in the bioreactor, S, is much greater than the saturation constant, then the biomass specific growth rate equals the maximum specific growth rate and therefore the substrate is not rate limiting. If, on the other hand, the substrate concentration is less than the saturation constant, then the specific growth rate drops below the maximum and the substrate becomes rate limiting.

The Monod equation is analogous to the Michaelis-Menten equation for enzyme kinetics. Like the Michaelis-Menten kinetics, a double reciprocal plot gives estimates of Monod kinetic parameters.

$$\frac{1}{\mu} = \frac{1}{\mu_x} + \frac{K_x}{\mu_x} \frac{1}{S} \tag{6-7d}$$

In a laboratory or pilot scale bioreactor, biomass and substrate concentration measurements can be taken during the early lag phase and early exponential growth phase of the fermentation.

This data can then be used to generate a plot of $1/\mu$ versus $1/S$. This is referred to as a Lineweaver-Burk plot. The slope is K_x/μ_x and the y-intercept is $1/\mu_x$. The x-intercept is $-1/K_x$. Shuler et al. [16] caution that the curve does not necessarily give good estimates of K_x. A plot based on the reciprocal of data points is not suitable for regression analysis.

A laboratory scale, continuous stirred tank reactor provides another way to determine the biomass maximum specific growth rate. If the substrate concentration in the feed is much greater than the saturation constant, then the dilution rate approaches the specific growth rate. As the feed dilution rate is increased, the specific growth rate of the microbe increases until it reaches a maximum specific growth rate. At this point, the cells are

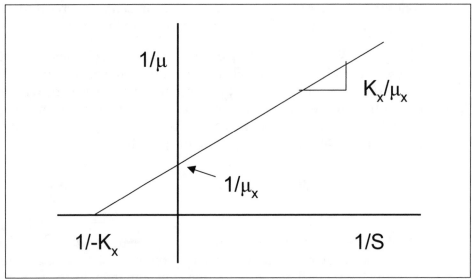

Figure 6-7a. A Double-Reciprocal or Lineweaver-Burk Plot.

washed out of the bioreactor. The dilution rate at washout is approximately the biomass maximum specific growth rate.

The relationship between substrate and specific growth rate used by Birol et al. [6] is a slight modification of Monod kinetics that is referred to as the Contois kinetics:

$$\mu = \frac{\mu_x S}{K_x X + S} \tag{6-7e}$$

Birol et al. [6] explain that biomass growth is known to be inhibited by high biomass concentrations in penicillin production bioreactors. Contois kinetics describes substrate limitation at high cell density, and as substrate concentration decreases the specific growth rate is inversely proportional to biomass concentration.

An important component of the unstructured mode proposed by Birol et al. [6] is that the effects of pH, temperature, and dissolved oxygen on the biomass specific growth rate are considered.

Including these variables increases the utility of the model in optimization and control studies.

Contois kinetics are used to describe the effect of dissolved oxygen:

$$\mu \propto \frac{C_L}{K_{ox} X + C_L} \tag{6-7f}$$

C_L is the dissolved oxygen concentration, and K_{ox} is the dissolved oxygen saturation constant.

The Birol et al [6] equation to calculate the concentration of hydrogen ions in the culture broth is in charge balances, section 6.5 of this chapter. Birol et al. [6] use a simple model given by Nielsen et al. [5] to describe the effect of pH on biomass growth as shown in equation 6-7g.

$$\mu \propto \left[\frac{\mu_x}{1 + \dfrac{K_1}{[H^+]} + \dfrac{[H^+]}{K_2}} \right] \tag{6-7g}$$

Biomass growth rate increases with increasing temperature to a maximum and decreases at culture temperatures beyond this maximum. Shuler et al. [16] describe this behavior as the difference between Arrhenius equations for thermal growth and death (equation 6-7h).

$$\mu \propto \left\{ \left[k_g \exp\left(-\frac{E_g}{RT} \right) \right] - \left[k_d \exp\left(-\frac{E_d}{RT} \right) \right] \right\} \tag{6-7h}$$

The optimum temperature for biomass growth depends on culture pH. Since the culture temperature is held constant during a production run and under the assumption of homogeneous conditions in the stirred tank bioreactors, the effects of pH and temperature on one another are ignored.

The overall equation that describes specific growth rate is the product of all these effects as in the following equation:

$$\mu = \frac{\mu_x S}{K_x X + S} \frac{C_L}{K_{ox} X + C_L} \left[\frac{\mu_x}{1 + \dfrac{K_1}{[H^+]} + \dfrac{[H^+]}{K_2}} \right] \tag{6-7i}$$

$$\times \left\{ \left[k_g \exp\left(-\frac{E_g}{RT} \right) \right] - \left[k_d \exp\left(-\frac{E_d}{RT} \right) \right] \right\}$$

A mass balance on penicillin formation is similar to equation 6-7b, describing biomass growth.

Some denaturation of product is expected during the course of the fermentation. Product loss is directly related to the product formation by a hydrolysis constant, K.

$$\frac{dP}{dt} = \mu_{pp} X - KP - \frac{P}{V}\frac{dV}{dt}$$ (6-7j)

Like the specific biomass growth rate, the specific product formation rate is related to a maximum specific growth rate. Instead of Monod or Contois kinetics, Birol et al. [6] use noncompetitive substrate inhibition kinetics to describe the dependence of product formation on substrate concentration in the bioreactor. This equation was chosen because of its empirical fit and not because it describes a physiological state of penicillin production.

$$\mu_{PP} = \mu_P \frac{S}{K_P + S + S^2 / K_I}$$ (6-7k)

Shuler et al. [16] place a $K_I \gg K_p$ requirement on this equation. Birol et al. [6] set $K_P = 0.0002 Kg/m^3$ and $K_I = 0.1 Kg/m^3$.

The effect of dissolved oxygen concentration on product rate, just as with biomass growth rate, is described by Contois kinetics. The overall expression for specific penicillin formation rate is as follows:

$$\mu_{PP} = \mu_P \frac{S}{K_P + S + S^2 / K_I} \frac{C_L^p}{K_{op} X + C_L^p}$$ (6-7l)

where the p in C_L^p is an exponent, not a descriptor.

All three bioreactor simulations use glucose or a polymer of glucose as a growth-limiting substrate. The portion of the glucose consumed for energy must be included in a mass balance. Birol et al. [6] relate substrate consumption for energy directly to biomass concentration using a maintenance coefficient, m_x. Biomass and product yields on substrate, $Y_{X/S}$ and $Y_{P/S}$, respectively, are assumed to be constant throughout the fermentation.

$$\frac{dS}{dt} = -\frac{\mu}{Y_{X/S}} X - \frac{\mu_{PP}}{Y_{P/S}} X - m_x X + \frac{Fs_f}{V} - \frac{S}{V}\frac{dV}{dt}$$ (6-7m)

Where S_f is the concentration of substrate in the feed stream.

Human Epidermal Growth Factor Production

This simulation describes the effect of the metabolic burden on a host cell as a result of the expression of a recombinant molecule from a plasmid cloning vector. Naturally occurring plasmids are small nongenomic, circular strands of DNA found in bacteria and some yeast and fungi. Some types of plasmids can replicate independently of the genome, and conjugative plasmids can transfer between bacteria of the same species.

Plasmids can confer antibiotic resistance or a competitive advantage to bacteria that contain them. R factor plasmids contain genes that code for enzymes capable of destroying or modifying antibiotics. Since they are conjugative, R factors can confer antibiotic resistance on a whole population [11].

Plasmids are useful as cloning vectors when no posttranscriptional modifications to the protein are necessary. The bacteria host lacks the ability to make enzymatic changes to a protein after it is synthesized. Foreign DNA can be added to a plasmid along with an inducer to initiate expression of the foreign protein.

Bacteria like *Escherichia coli* do not normally excrete protein. A foreign protein that remains in the intracellular space is subject to degradation by proteolytic enzymes. The foreign proteins can also become insoluble. A signal or leader sequence can be added to enable secretion of the protein across the plasma membrane. The foreign protein then accumulates in the periplasmic space between the plasma membrane and the outer membrane or cell wall.

The disruption of cell function caused by the accumulation of foreign protein and the consumption of resources for the purpose of expressing a foreign protein that confers no advantage to the cell leads to the genetic instability of recombinant bacteria.

A single bacterium can contain as many as 40 copies of a plasmid [11]. These multi-copy plasmids are usually distributed randomly in daughter cells. There is a probability that daughter cells contain no plasmids. Host cell mutations and the structural alteration of plasmids can reduce or stop foreign protein expression [17].

To model this genetic instability in bacteria, the model by Zheng et al. [8] segregates the culture into plasmid-bearing, plasmid-free, and viable but nonculturable cells. The model contains a probability that daughter cells will become plasmid free, and it estimates extracellular protein produced by cell lysis. Note that this extracellular protein is not the foreign protein of commercial interest.

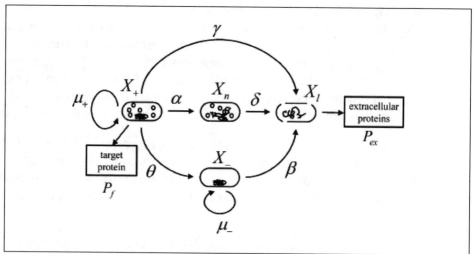

Figure 6-7b. Segregated Kinetic Model of Recombinant *E. coli* from Zheng et al. [8].

Biomass is divided into the three cell types: the plasmid-bearing cells, the plasmid-free cells, and the viable but nonculturable cells.

$$X = X_+ + X_- + X_n \tag{6-7n}$$

The mass balances on the three types of biomass are as follows:

$$\frac{dX_+}{dt} = (\mu_+ - \theta\mu_+ - \alpha - \gamma)X_+ - \frac{X_+}{V}F \tag{6-7o}$$

$$\frac{dX_-}{dt} = (\mu_- - \beta)X_- + \theta\mu_+ X_+ - \frac{X_-}{V}F \tag{6-7p}$$

$$\frac{dX_n}{dt} = \alpha X_+ - \delta X_n \frac{X_n}{V}F \tag{6-7q}$$

$$\frac{dX_l}{dt} = \beta X_- + \gamma X_+ + \delta X_n - \frac{X_l}{V}F \tag{6-7r}$$

X_l represents the loss of cell mass due to lysis. The amount of protein introduced into the culture medium by cell lysis is a fixed percentage of the lost cell mass.

$$P_{ex} = \varphi X_l \tag{6-7s}$$

Monod kinetics describe the specific growth rates of both plasmid-bearing and plasmid-free cells. The saturation constants are the same for both cell types. The maximum specific growth rate and saturation constants are the same for both cell lines. The maximum specific growth rate of plasmid-bearing cells is only slightly smaller than that of the plasmid-free cells. Before induction the culture grows at nearly the same rate as an unsegregated culture. The specific growth rate of plasmid bearing cells is modified by an induction factor that decreases the specific growth rate as the concentration of the intracellular inducer increases.

$$\mu_+ = \frac{\mu_{m+}S}{K_m + S}\left(1 - \frac{k_1 I_i}{k_2 + I_i}\right) \tag{6-7t}$$

$$\mu_- = \frac{\mu_{m-}S}{K_m + S} \tag{6-7u}$$

The effects of dissolved oxygen, pH, and temperature on specific growth rate, described in the previous section, are included in both specific growth rate calculations. These environmental variables are assumed to affect both cell types equally.

The inducer is injected into the bioreactor at some time after the fed-batch phase has begun. Zheng et al. [8] assume extracellular inducer concentration to be constant, since the cell volume is much smaller than the culture volume.

A change of intracellular inducer concentration has a first-order dependence on extracellular inducer concentration. The rate is decreased by the inducer already taken up.

$$\frac{dI_i}{dt} = D\left(\frac{I_e}{E + I_e}\right)\mu_{m+} - \mu_{m+}I_i \tag{6-7v}$$

The amount of intracellular inducer also affects the rate of conversion of plasmid-bearing cells, X_+, to viable but nonculturable cells, X_n. Monod kinetics describes the conversion rate as in the following equation:

$$\alpha = \frac{AI_i}{B + I_i} + C \tag{6-7w}$$

Target product formation is described as a foreign protein. The mass balance is identical to that used in the penicillin production model. The specific foreign protein production rate, π, is based on Monod kinetics of inducer transport into the cell.

$$\frac{dP_f}{dt} = \pi X_+ - k_5 P_f - \frac{P_f}{V} F \qquad (6\text{-}7x)$$

$$\pi = k_3 \mu_+ \frac{I_i}{K_\Pi + I_i} + k_4 \qquad (6\text{-}7y)$$

The substrate utilization model assumes that all three cell types have the same maintenance rates.

$$\frac{dS}{dt} = -(\mu_+ X_+ + \mu X_-)/Y_{X/S} - m_x X + \frac{FS_f}{V} - \frac{S}{V}\frac{dV}{dt} \qquad (6\text{-}7z)$$

Fuel Ethanol Production by Simultaneous Saccharification and Fermentation

The enzyme hydrolysis of a polysaccharide-like starch or cellulose to glucose is strongly inhibited by glucose. Simultaneous saccharafication and fermentation reduces this inhibition by fermenting the glucose in the same reaction vessel. Mosier and Ladisch [7] have modeled the hydrolysis of cellulose to glucose and the subsequent fermentation of the glucose to ethanol.

Although multiple enzyme groups work in concert to solubilize and hydrolyze cellulose to glucose, Mosier and Ladisch [7] model the cellulose hydrolysis with two reactions. Cellulose is hydrolyzed to cellobiose by exoglucanase with cellobiose inhibition, and then cellobiose is converted to glucose by β-glucosidase with glucose inhibition. The Michaelis-Menten equation with substrate inhibition governs both reactions as shown in equation 6-7aa.

$$\frac{dC}{dt} = \frac{-V_m C}{K_m \left(1 + \dfrac{C}{K_m} + \dfrac{S}{K_i} \right)} \qquad (6\text{-}7aa)$$

In the first reaction, C represents cellulose, and S represents cellobiose. In the second reaction, C represents cellobiose, and S represents glucose.

Ethanol is a by-product of yeast fermentation and is directly related to biomass concentration.

Mosier and Ladisch describe the mass balance on biomass in terms of the product formation.

$$\frac{dX}{dt} = XE\nu \qquad\qquad (6\text{-}7\text{ab})$$

$$E\nu = \mu \qquad\qquad (6\text{-}7\text{ac})$$

E is the portion of substrate utilized for biomass. ν is the specific ethanol production rate.

Ethanol noncompetitively inhibits yeast biomass growth at concentrations above 5 percent. Shuler et al. [16] and Mosier and Ladish [7] both use a common rate expression for ethanol inhibition.

$$\nu = \nu_m \left(\frac{S}{K_s + S} \right)\left(1 - \frac{P}{P_m} \right)^n \qquad\qquad (6\text{-}7\text{ad})$$

This simulation adds the equations for temperature and pH effects on biomass growth given in Birol et al. [6-6] to the ethanol production rate expression.

The ethanol mass balance is simple.

$$\frac{dP}{dt} = \nu X \qquad\qquad (6\text{-}7\text{ae})$$

The glucose mass balance appears similar to a fed-batch substrate mass balance; cellulose hydrolysis rather than a feed stream provides the glucose input.

$$\frac{dS}{dt} = -\frac{\mu}{Y_{X/S}} X + \frac{dC}{dt} \qquad\qquad (6\text{-}7\text{af})$$

The production of ethanol from corn starch by SSF has also been studied. The enzyme hydrolysis of starch to fermentable glucose is modeled differently from cellulose hydrolysis in that sugars of different degrees of polymerization, or DP, are hydrolyzed to glucose in parallel rather than series reactions.

Hybridoma Growth and Monoclonal Antibody Production
Bacteria plasmids were the first cloning vectors, and they are the easiest to work with. However, a plasmid vector cannot contain a large foreign

DNA molecule. Also, bacteria lack machinery to modify a protein after translation [11]. These posttranslational modifications include adding carbohydrates to form a glycoprotein or folding the protein into a three-dimensional structure. Glycoproteins are often located at the cell membrane, with a portion of them located in the extracellular space to act as signaling agents. Many proteins must be folded into a three-dimensional structure, also called its tertiary structure, in order to function properly [1].

Growing mammalian cells in stirred tank bioreactors is difficult and expensive. The culture medium is complex, the cells are sensitive to shear, and the growth rate is slow relative to bacteria, yeast, or fungi. However, mammalian cells can produce glycoproteins like antibodies.

Antibodies are secreted by B lymphocyte cells into the blood in response to an antigen. An antigen is typically a protein, but can be any molecule that induces a B lymphocyte to produce an antibody. Large quantities of antibodies cannot be produced by culturing B lymphocyte cells because the cells can reproduce only a limited number of times. This limitation is overcome by fusing two types of B lymphocytes. One type is producing an antibody in response to antigens. The other type is an "immortal" B lymphocyte tumor. An immortal cell can reproduce indefinitely in culture. Some of the cells that result from the fusing are hybrid. They produce a single type of antibody or monoclonal antibody, and they reproduce indefinitely in culture. These clones are called *hybridomas*.

Monoclonal antibodies are produced from an immortal cell line called a hybridoma.

The structured model of hybridoma growth proposed by Batt and Kompala [14] contains three primary substrates. Mammalian cells utilize glucose as a carbon and energy source much as do microbes. The amino acid glutamine is the second primary substrate. Glutamine is the primary source of nitrogen but also acts as a carbon and energy source. Unlike microbes, hybridomas cannot manufacture some amino acids. These essentially are the third primary substrate.

The hybridoma excretes lactate from glucose metabolism and lactate and ammonia from glutamine metabolism. The monoclonal antibody is excreted by the hybridoma and is not included in the cell mass. This contrasts with the accumulation of foreign protein in recombinant bacteria.

Rather than a single, uniform biomass, the cells are divided into four parts referred to as pools: nucleotides, lipids, amino acids, and proteins. Each pool is defined in the model by a separate mass balance equation with inputs and outputs from substrates, products, and other pools. For instance, the mass balance for the nucleotide pool contains contributions from the glucose and glutamine substrates as well as from the amino acid pool. The mass balance on the amino acid pool has inputs from glucose, glutamine, and the amino acids in the medium; outputs to the protein, lipid, and nucleotide pools; and an output to the monoclonal antibody product.

Increased complexity in unstructured models is evidenced by the twelve component mass balances that are needed to define this simple structured model. Each substrate and product is defined by separate mass balance. Each pool has a separate mass balance, and separate mass balances are given for viable and nonviable cells.

The specific growth rate of the culture must be derived from the intracellular pool material balances. Since the sum of the overall cell mass fractions is always unity, the time differential is zero.

$$\frac{d}{dt}\left(\sum_{i=1}^{4}C_i\right) = \frac{d}{dt}(A+N+P+F) = 0 \tag{6-7ag}$$

A, N, P, and F represent the amino acid, nucleic acid, protein, and lipid pools, respectively. C_i represents the mass fraction of the ith pool.

The sum of the material balances for each pool is then used to define the specific growth rate in terms of synthesis and depletion terms for each pool, r_{ij}.

$$\sum_{i=1}^{4}\left[\frac{dC_i}{dt}\right] = \sum_{i=1}^{4}\sum_{i=j}^{4}r_{ij} - \mu\sum_{i=1}^{4}C_i \tag{6-7ah}$$

$$\mu = \sum_{i=1}^{4}\sum_{j=1}^{4}r_{ij} \tag{6-7ai}$$

Substituting the original rate expression gives a specific growth rate in terms of all intracellular pools and monoclonal antibody production, I_g.

$$\mu = \left[\frac{dA}{dt}\right]_s + (1-a_{AN})\left[\frac{dN}{dt}\right]_s + (1-a_{AP})\left[\frac{dP}{dt}\right]_s + (1-a_{AF})\left[\frac{dF}{dt}\right]_s - a_{AI}\left[\frac{dI_g}{dt}\right]_s \tag{6-7aj}$$

6-8.   Mass Transfer

A primary concern when designing and operating an aerated bioreactor is moving sufficient oxygen for the cells from the gas phase to the liquid phase. Carbon dioxide is not a substrate in the production of penicillin or hEGF, but knowledge about carbon dioxide mass transfer from the liquid to the gas phase is useful when analyzing off-gas balance data.

Oxygen Interphase Mass Transfer

Oxygen is sparingly soluble in water. At 25°C it is 1.26 mmole/L. A typical, aerated bioreactor consumes 750 times the saturation volume per hour [16].

> *Oxygen transfer from the sparged gas to the culture media is a*
> *key parameter in the design and operation of a bioreactor.*

Measuring the concentration of oxygen at the gas-liquid interface is not practical. Oxygen concentration in gas phase can be measured directly. Dissolved oxygen probes in the bulk liquid can be problematic, but a measurement can be made.

The driving force is the difference between the bulk liquid-phase oxygen concentration in equilibrium with the bulk gas-phase oxygen concentration, as given by Henry's law.

$$C_G = HC_L^*$$ (6-8a)

C_G is the concentration of oxygen in the exit gas, C_L^* is the concentration of oxygen in equilibrium with the exit gas, and H is Henry's law constant [16].

Another approximation is that all the resistance to mass transfer lies on the liquid side. The flux of oxygen to the liquid phase at steady state is then

$$\text{Oxygen Flux} = k_l\left(C_l^* - C_l\right)$$ (6-8b)

k_l is the oxygen transfer coefficient.

The oxygen absorption rate depends on the area of the gas-liquid interface per unit volume of liquid in the bioreactor. This parameter, a, is often included with the oxygen transfer coefficient to become the volumetric oxygen transfer coefficient, $k_l a$. The gas absorption rate, Q_{O_2}, is then

$$Q_{O_2} = k_l a\left(C_l^* - C_l\right)$$ (6-8c)

An important implication of this relationship is that the oxygen transfer rate is equivalent to the oxygen uptake rate of the microbes [17].

Estimating Volumetric Oxygen Mass Transfer Coefficient in a Bioreactor

$k_l a$ is estimated in an operating bioreactor by momentarily shutting off the air and measuring the rate at which dissolved oxygen decreases. A mass balance on the oxygen concentration in the liquid phase can be summarized as follows:

$$\frac{dC_l}{dt} = k_l a\left(C_l^* - C_l\right) - q_{O_2} X \tag{6-8d}$$

q_{O_2} is the specific oxygen uptake rate. $-q_{O_2} X$ is the total oxygen uptake rate and is referred to by the acronym OUR.

When the air is turned off, the $k_l a(C_l^* - C_l)$ term falls out, and the slope of the decreasing dissolved oxygen concentration is equal to $-q_{O_2} X$. Remember that C_l^* is known from Henry's law and is calculated from the oxygen concentration in the exit gas. C_l is the dissolved oxygen concentration. The two sources of error in this measurement are the accuracy of the biomass measurement and the time constant of the dissolved oxygen electrode. Dissolved oxygen can be allowed to drop by only a certain amount before the culture is affected. If the oxygen uptake rate is too high and this point is reached quickly, the time constant of the dissolved oxygen probe can introduce errors into the $k_l a$ measurement.

The estimation of $k_l a$ by using this gassing-out method can be automated to run periodically throughout a batch. A virtual plant can then be updated with the new $k_l a$ value.

Power Requirement versus Volumetric Oxygen Mass Transfer Coefficient

Mechanical agitation in an aerated bioreactor reduces bubble size and increases the gas-liquid interfacial area. The volumetric oxygen mass transfer coefficient can be correlated to the power per-unit volume. This P/V ratio is one of several criteria for bioreactor scale up [18], and is an indirect way to scale oxygen transfer.

Bioreactor mixing properties include tip speed, power-per-unit volume, Reynolds number, and pumping capacity per unit volume. Not all of these properties can be kept constant on scale up.

In the case of simulations, the roll of this correlation is reversed. Known values of k_la are used to calculate power requirements. Birol et al. [6] employ a correlation given in Bailey and Ollis [15], but this equation is scaled for single impeller systems that have laboratory or pilot scale working volumes. An industrial scale correlation by Fukuda et al. [19] and quoted in Kargi and Moo-Young [20] is used in both the penicillin and the hEGF production simulations.

The Fukuda et al. [19] correlation is suitable for 100-L to 42,000-L working volumes. An important feature of this correlation for industrial scale bioreactors is an additional parameter, N_i, that allows for multiple impellers.

$$k_l a = \left(\chi + \delta N_i \right) \left(P_G / V_L \right)^{0.77} V_S^{0.67} \qquad (6\text{-}8e)$$

V_s is the superficial velocity of the gas bubbles; P_G/V_L is the power input per unit volume of the bioreactor.

Carbon Dioxide Interphase Mass Transfer
Yaqi and Yoshida [21] suggest that the partial pressure of carbon dioxide in a bioreactor exit gas is, for practical purposes, in equilibrium with the dissolved carbon dioxide concentration. Bailey and Ollis [15] and Nielsen et al. [5] point out that dissolved CO_2 exists in an equilibrium with H_2CO_3, HCO_3^-, and CO_3^{2-} that is sensitive to pH. The rate of removal of carbon dioxide from the gas phase depends on an unfavorable chemical equilibrium as well as mass transfer resistance.

Oxygen and Carbon Dioxide Mass Balances in Penicillin Production
The dissolved oxygen mass balance equation is similar to the substrate balance. All the mass balances are based on the liquid phase in the bioreactor, so the dissolved oxygen mass balance must include the gas-liquid mass transfer of oxygen.

$$\frac{dC_L}{dt} = -\frac{\mu}{Y_{X/O}} - \frac{\mu_{PP}}{Y_{P/O}} - m_o X + k_L a \left(C_L^* - C_L \right) - \frac{C_L}{V} \frac{dV}{dt} \qquad (6\text{-}8f)$$

In the face of the complexity of a pH sensitive equilibrium and an apparent lack of carbon dioxide mass transfer data, Birol et al. [6] chose to model the carbon dioxide evolution with a purely empirical equation.

$$\frac{dCO_2}{dt} = \alpha_1 \frac{dX}{dt} + \alpha_2 X + \alpha_3 \qquad (6\text{-}8g)$$

Oxygen and Carbon Dioxide Mass Balances in hEGF Production
Roeva et al. [12] estimated the parameter values used in the dissolved
oxygen and carbon dioxide mass balances. The mass balance on dissolved
oxygen is similar to that used in penicillin production except that biomass
growth is segregated into plasmid-bearing and plasmid-free.

$$\frac{dC_L}{dt} = -(\mu_+ X_+ + \mu X_-)/Y_{DO/X} - m_o X + k_L^{O_2} a(C_L^* - C_L) - \frac{C_L}{V}\frac{dV}{dt} \qquad (6\text{-}8h)$$

Since Roeva et al. [12] provide a mass transfer coefficient for carbon
dioxide, $k_L^{CO_2}a$, a mass transfer equation is used in the hEGF simulation
rather than the empirical equation used that was in the penicillin
production simulation.

$$\frac{dCO_2}{dt} = -(\mu_+ X_+ + \mu X_-)/Y_{CO_2/X} - m_{CO_2} X + k^{CO_2}{}_L a(CO_2^* - CO_2) - \frac{CO_2}{V}\frac{dV}{dt} \qquad (6\text{-}8i)$$

Estimates of the maintenance and product formation parameters in the
hEGF production simulation are taken from Birol et al. [6].

Gas Liquid Mass Transfer in a Mammalian Cell Culture
One of the fundamental limitations to gas-liquid mass transfer of oxygen
in a mammalian cell culture bioreactor is the shear sensitivity of the cells.
Microbes have cell membranes and cell walls designed to protect them in
a natural environment. Mammalian cells bind together to form tissues and
adhere to a solid surface. Early industrial cell culture reactors were roller
bottles.

Submerged culture in an aerated, stirred tank bioreactor was itself a
significant technical undertaking.

Agitating aerated bioreactors reduces bubble size and increases the $k_l a$
term in the coefficient. The smaller bubble size also increases their
residence time in the bioreactor. Because of the shear sensitivity of cell
cultures, the power-per-unit volume is 100 times less than the power-per-
unit volume of a microbial bioreactor. Adding more air to the bioreactor to
compensate for the low P/V ratio is not feasible either. Bubbles bursting at
the liquid surface can break up high numbers of mammalian cells, and
amino acids and proteins in the culture medium can cause foaming
problems at high gas flow rates.

To minimize shear, the gas flow rate to the sparger and the stirring rate
can be kept constant.

Dissolved oxygen can be controlled by mixing pure oxygen with the air to the sparger. An alternate configuration is to supply air or oxygen only to the head space. When bubble aeration is used, the ratios of air, oxygen, carbon dioxide, and inert gas can be maintained when the head space is diluted with an inert gas like helium.

Measuring the bubble gas phase mass transfer coefficient is more difficult on slow growing mammalian cell bioreactors because of the oxygen transfer from the head space. Oeggerli et al. [22] give a liquid phase oxygen mass balance in a bioreactor with separate bubble and headspace gas phases (as shown in equation 6-8j).

$$V_L\left(\frac{dC_L}{dt}\right) = k_l a_B\left(C_{LB}^* - C_L\right)V_L + k_l a_H\left(C_{LH}^* - C_L\right)V_L + OUR \qquad \text{(6-8j)}$$

The subscript H in this equation represents head space, L represents the bioreactor culture medium, and B represents the bubble phase. When the gas feed to the bubble phase is turned off, $k_l a_B$ drops to zero. The mass balance is left with the head space mass transfer coefficient.

$$OUT = V_L\left(\frac{dC_L}{dt}\right) - k_l a_H\left(C_{LH}^* - C_L\right)V \qquad \text{(6-8k)}$$

The head space mass transfer coefficient, $k_l a_H$, and the mole fraction of oxygen in the head space may vary during the cultivation. Variations in the mole fraction of oxygen affect the calculation of the equilibrium concentration, C_{LH}^*. Oeggerli et al. [22] suggest that errors in these estimates can be reduced by maintaining a constant agitation rate and keeping the head space well mixed.

Evaporation Loss from a Fed-Batch Bioreactor

Birol et al. [6] give an estimated loss of broth of 10 percent to 20 percent in an industrial bioreactor. The evaporation is simply the humidity picked up by the relatively dry gas entering the bioreactor as it bubbles through. Birol et al. [6] provide a heuristic estimate of evaporative losses that depends on the freezing point, boiling point, and volume of the broth.

$$F_{loss} = V\lambda\left(\exp\left(5\frac{T - T_0}{T_v - T_0}\right) - 1\right) \qquad \text{(6-8l)}$$

This estimate is used in all three simulations.

6-9.   Simulated Batch Profiles

In this section the dynamic process models developed in previous sections are running on a distributed control system. These virtual plants all run faster than real time, and the models' biomass, liquid, dissolved, and gas phases can be sped up individually to maintain integration stability for vastly different rates of change.

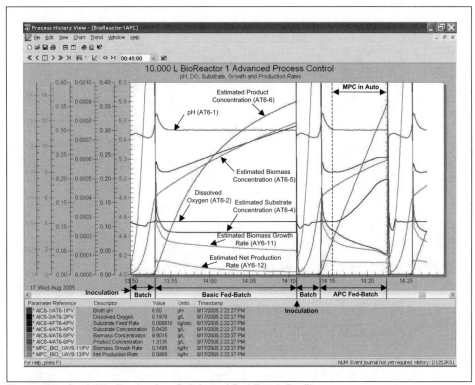

Figure 6-9a. Penicillin Production, Batch, and Fed-Batch Profiles

The batch profile of penicillin production shows both batch and fed-batch phases. The dissolved oxygen is controlled by the split-range manipulation of air flow and head pressure, and the substrate concentration is controlled by the manipulation of the glucose feed rate. The initial glucose concentration was 15 Kg/m^3; the fed-batch phase was initiated when it fell to 0.1 Kg/m^3.

A higher dissolved oxygen concentration progressively increased the biomass growth and product formation rates as the cell concentration increased. High air flow increases gas processing costs, and high pressures increase the dissolved carbon dioxide in the broth. As a result, the dissolved oxygen was kept lower in the beginning of the batch when the beneficial effect was minimal. A lower substrate concentration decreased the biomass growth but increased the product formation rates because of a reduced substrate inhibition factor. Consequently, the high substrate concentration was initially high to grow cells but was kept low in the fed-batch phase to promote product formation.

The diagram also shows a subsequent fed-batch profile that has the same simulation and initial conditions. The MPC was automatically turned on at a minimal product concentration and turned off just before the end point.

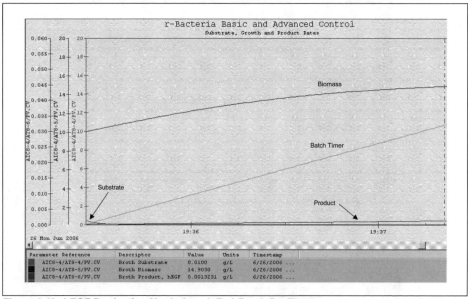

Figure 6-9b. hEGF Production Not Induced, Fed-Batch Profile

Figures 6-9b and 6-9c show the fed-batch portions of two fermentations. The initial conditions and fed-batch phase length are the same, but an inducer was not added to the run profiled in figure 6-9b. Without induction, very little hEGF was produced, and the biomass reached 14.9 Kg/m^3. In the profile shown in figure 6-9c, inducer was added six hours after the start of the fed-batch phase. The adverse effect of hEGF production on the cells is evident from the drop in final biomass concentration.

Figure 6-9d shows three reactions: hydrolysis of cellulose to cellobiose, hydrolysis of cellobiose to glucose, and production of ethanol and yeast from glucose. The charts displays a batch that is run 100 times real time. The batch cycle time of about 30 hours and the ethanol end point of about 48 g/L are consistent with published results. Note the low concentrations of the inhibitory cellobiose and glucose.

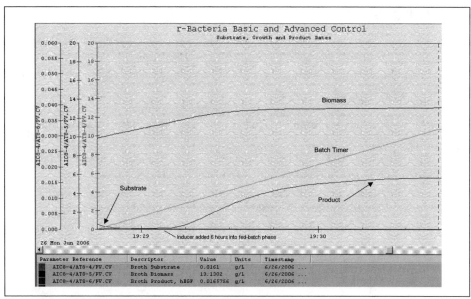

Figure 6-9c. hEGF Production Induced, Fed-Batch Profile

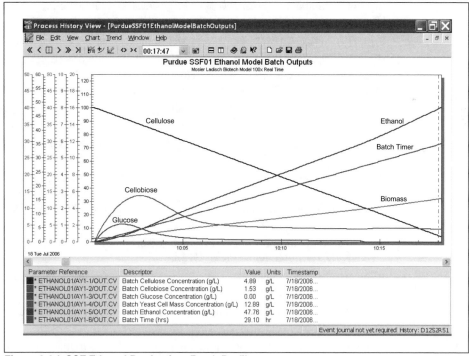

Figure 6-9d. SSF Ethanol Production, Batch Profiles

References

6-1.   Alberts, Bruce, Johnson, Alexander, Lewis, Julian, Raff, Martin, Roberts, Keith, and Walter, Peter, *Molecular Biology of the Cell*. 4th ed. NY: Garland Science, 2002.

6-2.   Beckerman, Martin, *Molecular and Cellular Signaling*. New York: Springer-Verlag, 2005.

6-3.   Tomita, M. , Hashimoti, K., Takahashi, K., Shimizu, T. S., Matsuzakei, Y., Miyoshi, F., Saito, K., Tanida, S., Yugi, K., and Hutchison Venter III, J. C., "E-CELL: Software Environment for Whole-cell Simulation." *Bioinformatics*, 15 (1999): 72-84.

6-4.   Seborg, Dale E., Edgar, Thomas F., and Mellichamp, Duncan A., *Process Dynamics and Control*. 2d ed. Hoboken, NJ: Wiley. 2004.

6-5.   Nielsen, Jens, Villadsen, John, and Liden, Gunnar, *Bioreaction Engineering Principles*. 2d ed. New York: Kluwer Academic/Plenum Publishers, 2003.

6-6.   Birol, Gulnur, Undey, Cenk, and Cinar, Ali, "A Modular Simulation Package for Fed-batch Fermentation: Penicillin Production." *Computers and Chemical Engineering*, 26 (2002): 1553-65.

6-7.   Mosier, N. S., and Ladisch, M. R., "Biotechnology." Unpublished lecture notes, 2005.

6-8.   Zheng, Zhi-Yong, Yao, Shan-Jing, and Lin, Dong-Qiang. "Using a Kinetic Model that Considers Cell Segregation to Optimize hEGF Expression in Fed-batch Cultures of Recombinant *Escherichia coli*." *Bioprocess Biosyst Engineering*, 27 (2005): 143-52.

6-9.   Britt, H., Chen, C-C., Mahalec, V., and McBrien, A., *Modeling and Simulation in 2004: An Industrial Perspective*. Cambridge, MA: Aspen Technology, Inc., 2004.

6-10.   McMillan, Gregory K., and Cameron, Robert A., *Advanced pH Measurement and Control*. 3d ed. Research Triangle Park, NC: ISA, 2005.

6-11.   Prescott, Lansing M., Harley, John P., and Klein, Donald A., *Microbiology*, 6th ed. Boston: McGraw-Hill, 2005.

6-12.   Roeva, Olympia and Tzonkov, Stoyan, "Modelling of *Escherichia coli* Cultivations: Acetate Inhibition in a Fed-Batch Culture." *Bioautomation*, 4 (2006): 1-11.

6-13.   Batt, Brian C., and Kompala, Dhinakar S., "A Structured Kinetic Modeling Framework for the Dynamics of Hybridoma Growth and Monoclonal Antibody Production in Continuous Suspension Cultures." *Biotechnology and Bioengineering*, 34: 515-31.

6-14.   Aiba, Shuichi, Humphrey, Arthur E., and Millis, Nancy F., *Biochemical Engineering*. 2d ed. New York: Academic Press, 1973.

6-15.   Bailey, James E., and Ollis, David F., *Biochemical Engineering Fundamentals*. 2d ed. New York: McGraw-Hill, 1986.

6-16.   Shuler, Michael L., and Kargi, Fikret, *Bioprocess Engineering Basic Concepts*. 2d ed. Prentice Hall, 2002.

6-17.   Dunn, Irving J., Heinzle, Elmar, Ingham, John, and Prenosil, Juri E., *Biological Reaction Engineering*. 2d ed. Wiley-VCH Verlag, Weinheim KgaA, 2003.

6-18.   Oldshue, James Y., *Fluid Mixing Technology*. New York: McGraw-Hill, 1983.

6-19.   Fukuda, H., Sumino, Y., and Kansaki, T., "Scale-up of Fermenters. II. Modified Equations for Power Requirement." *Journal of Fermentation Technology*, 46: 838-45.

6-20.   Kargi, F., and Moo-Young, M., "Transport Phenomena in Bioprocesses." In *Comprehensive Biotechnology*, edited by Murray Moo-Young, 57-76. New York: Pergamon Press, 1986.

6-21.   Yaqi, Hideharu, and Yoshida, Fumitake, "Desorption of Carbon Dioxide from Fermentation Broth." *Biotechnology and Bioengineering*, 19, no. 6: 801-19.

6-22.   Oeggerli, A., Eyer, K., and Heinzle, E., "Online Gas Analysis in Animal Cell Cultivation: II. Methods for Oxygen Uptake Rate Estimation and Its Application to Controlled Feeding of Glutamine." *Biotechnology and Bioengineering*, 45: 54-62.

Chapter 7:
Neural Network Industrial Process Applications

Chapter 7

Neural Network Industrial Process Applications

Robert L. Heider, Adjunct Professor
Chemical Engineering Department
Washington University
St. Louis, MO

Mike May, Process Engineer
Haldor Topsoe, Inc.
Pasadena, TX

Ashish Mehta, Principal Software Engineer
Emerson Process Management
Austin, TX

7-1.   Introduction

Imagine that you are the master brewer for your neighborhood microbrewery! For an upcoming event, you are to make a large batch of beer. The typical fermentation cycle takes a few weeks, so you have to make a judgment call as to when the perfect brew is ready (and avoid the wrath of your patrons). Biochemistry teaches us that one of the critical indicators of brew quality is the diacetyl content. However, you need to take the brew sample to a chemical analyst to measure that content, as no viable physical sensors exist.

Now imagine that you can get a new software program that will provide a real-time measurement of the diacetyl content throughout the fermentation cycle. You can monitor the quality, make sure your beer is not getting "malflavors," and stop the batch as soon as it reaches the desired value. Such a software sensor is now possible thanks to neural network (N N) technology [20].

When neural networks came on the industrial scene in the 1980s, they were viewed as a panacea for many of the difficult or unsolvable control and measurement problems that plagued the industrial control community. However, when neural networks were applied to these problems, few successes occurred. Initially, this gave them a bad name, but today there are many well-documented neural network applications in the process industry [2] [14].

The technical literature contains examples of many different applications for a wide variety of neural network uses. Some of these are electrocardiogram (EKG) noise cancellation, speech recognition, and sonar signals processing. Neural networks have even been used to predict the stock market [6]. Neural networks function as a model that can be used to predict a process state that could be a difficult-to-measure variable or an analytical value, such as a lab-measured variable [3] [14].

Learning Objectives

A. Identify the typical neural network applications in the process industry.

B. Understand the basic concepts of neural networks and how they are trained.

C. Recognize the importance of time (dynamics) in industrial process applications.

D. Appreciate the requirement to generalize neural network–based models.

E. Know how to satisfy the data quality and quantity requirements for neural networks.

F. Learn how to develop a successful neural network application.

G. Gain a perspective via the anatomy of two neural network applications.

H. Be aware that neural networks are also used directly as controllers.

I. Look ahead to future developments and applications of the technology.

In the process industries, one frequent application of neural networks is as a "soft sensor." The concept is that if many physical measurements about the process are known then the value of an unmeasured variable can be computed, in real time, using a neural network software program. This unmeasured value can be an analytical measurement, such as the analysis of a power plant stack. The time delay in traditional lab analysis implied that the product must either be reworked or, in some processes, disposed of. By providing online measurements, soft sensors increase the efficiency and profitability of the process.

> *Neural networks can be designed to provide real-time measurements and validation of critical analytical properties. These networks are called "soft sensors."*

There are many documented applications of successful soft sensor applications. One is power plant stack emission. An analytical instrument is rented and installed on the plant stack. Next, a large amount of data is taken over several operational conditions. This data is then used to develop and train a neural network to model the analytical instrument.

When a process is well understood, robust models can be built based on the first principles that describe the process. These are called "white boxes." However, when a process has considerable complexity, the governing first principles may be unknown. In these situations, a heuristic model, such as a neural network, that is generated by correlating process data with laboratory analysis is practical because such models take less time to develop than do first-principle models. Neural networks are frequently called "black boxes." Even in situations where a deterministic model is available, heuristic models can be used to enhance the results because first-principle models have inherent inaccuracies due to approximations, noisy data, and unmeasured second-order effects. Two such cases where a heuristic model can be used to supplement a deterministic model are when closing a model mismatch and when accomplishing faster execution [17]. Closing a model mismatch entails training a neural network to minimize the error between the output from the equation-based model and the actual process data. For faster execution, a neural network model can be trained using data available from the equation-based model and executed at a higher frequency, which allows for faster problem solving.

Neural networks are heuristic models, based on discovery.

Another form or use of neural networks that is gaining acceptance is a combination of first principal or white boxes with the black box neural network. The resulting hybrid is called a "gray box." An example of this is bioprocess applications in which the equations governing cell growth are known, but some of the inputs are not known. Neural network outputs can then act as inputs to known first-principal equations [13].

Frequently, process plants are constructed in such a way that it is difficult to install some of the instruments needed to provide the necessary inputs to a deterministic process model. In these cases, there may be no proper piping or equipment configuration to provide an accurate and repeatable process measurement. For some of these situations, the value of the process variable may be inferred by an implied measurement. For example, as shown in figure 7-1a, crammed ducting will not allow sufficient distances for a flow meter, though a damper position can serve as a flow measurement approximation. For these instances, a neural network can also be developed that incorporates these implied values.

Figure 7-1a. Crammed Ducting: An Example of Difficult to Measure Process Variables

Just What Is a Neural Network?

Many excellent references describe neural network structures and their mathematics in detail [7]. As with other computational programs, neural networks have inputs and outputs; however, they process the data differently. Their operation is analogous to that of the fundamental cellular unit in the brain called the neuron. The human brain has as many as 100 billion neurons. They are often less than 100 microns in diameter and have as many as 10,000 connections, known as "axons," to other neurons. The neurons have input paths called "dendrites" that gather information from these axons. The connection between the dendrites and the axon is called a "synapse." The transmission of signals across the synapse is chemical in nature, and the magnitude of the signal depends on the amount of chemicals, called "neurotransmitters," that the axon releases. Many drugs work on the basis of changing these natural chemicals. The brain's memory process functions by combining the synaptic signal with the processing of information in the neuron [4] [5]. The term *artificial neural network* or ANN is frequently used to signify networks that are developed by programs or by wiring electronic circuits in contrast to physiological neural networks.

Neural networks, or more accurately, artificial neural networks, are computer programs that model the functionality of the brain. In artificial neural networks the processing element or PE is analogous to the biological neuron (figure 7-1b). Another term for the PE is "node." The PE has many inputs either from input data or other PE elements. Much like synaptic strength, the inputs are weighted and summed. The result of this summation is processed by a transfer function or a threshold function before it is passed to the output path. The output path is then passed to other inputs or to an output node. These PEs are organized in layers, which are grouped into input, output, and "hidden."

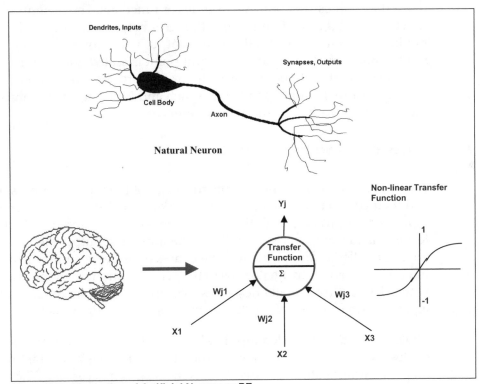

Figure 7-1b. Illustration of Artificial Neuron or PE

Computational neural networks are inspired
by the human brain.

Neural networks are different from artificial intelligence or AI systems. AI systems are rule based, and as a result a human expert is needed to work with the system to input the rules. Neural networks learn by processing the data without rules. Examples of known behavior are presented to the system, and computations adjust the system's junction weights to the required behavior. Obviously, someone knowledgeable in the simulated process is needed to input the data. The compilation of this data is called "learning" or "training" instead of programming. The network converges to a point. Testing this network against known data is called "recall." Because neural networks do not have rules, that is, are not rule based, they cannot explain how they arrived at a stated answer. There is a hidden danger in applying bad data to a system that has learned good behavior; it can unlearn. It has no way of knowing that the data it is learning with is bad or not as accurate or valid as the data it learned before.

Neural networks are not rule based.

Another way of describing a neural network is as a nonlinear regression. Input data values are applied to a series of functions. The weights of these functions are adjusted to minimize the error between the network's predicted output value and the actual output value corresponding to the input data. Each data input and output point is scaled, that is, it is normalized and centered on zero with a standard deviation of one. This is done to ensure that data points from each and every input and output to the neural network have the same reference.

7-2.   Types of Networks and Uses

A neural network is defined by the topology of the inputs, outputs, PEs, and their connection weights. The network is structured as layers, and the input and output layers are the points where data is entered and outputs leave. The PEs are layered, parallel to the input and output elements, in what are called *hidden* elements. The connections between the elements have varying strength, analogous to the synaptic strength in the brain. The network structure actually describes a complex matrix mathematical equation. An example of the "network" is shown in figure 7-2a.

The u values are the scaled input values. Y is the scaled output value. The circles are nodes. The lone node in the upper middle is a constant of one. Single-constant values are used to bias the results if necessary. The hidden layer PE of figure 7-2a processes the sum of the inputs through some function, such as a hyperbolic tangent as illustrated in figure 7-1b and in the following equation:

$$l_j = \sum_i w_{ij} x_i \qquad (7\text{-}2a)$$

The inner connecting lines correspond to weights or values, either positive or negative, which are multiplied by the input value or the intermediate node value. Solid lines are positive, and negative values are dotted. The sum of these weights multiplied by the input values (or hidden values) is then processed through a transfer function, commonly some nonlinear exponential function. In case the function is linear, the network becomes a linear regression. So for the example in figure 7-2a, a total of 19 weights have to be selected.

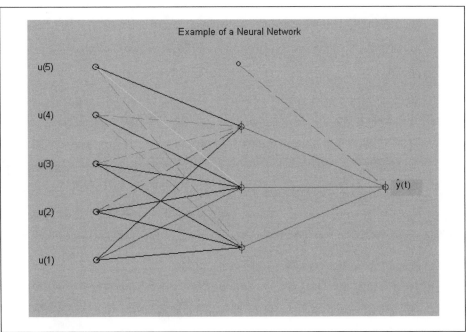

Figure 7-2a. Example of a Neural Network

Neural networks are structured in layers: one input, one output, and one or more hidden layers. The PEs or artificial neurons are in the hidden layer.

Consider a network in detail by studying a simple example, figure 7-2b. In this example, a neural network has been trained to model the outlet temperature of a shell-and-tube heat exchanger, acting as a cooler. Following the network from left to right, the network has three process inputs: the inlet fluid temperature, the inlet cold water flow rate, and the hot water flow rate. Each input is normalized around zero. One scaling method subtracts the mean value of the sample set from each input point and divides it by the standard deviation. The same method is used for the output value. The scaling factor result is presented as a scaled input to the input layer. In this example, an additional bias value of 1 is also present. Next is the hidden layer. Each PE is separated into two operations: the summation and the sigmoid transfer function. The lines from each input layer node to the summing junction node represent weights given to each input node. For clarity, not all weights are shown in figure 7-2b. The output of the summing junction acts as an input to the sigmoid transfer function. The outputs of this function are then factored by another set of weights as well as another bias term and then summed at the linear output layer. Finally, the output scaled value is rescaled to present the network output in engineering units.

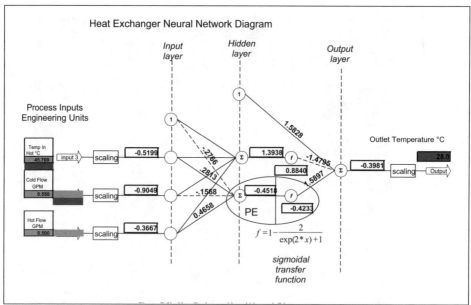

Figure 7-2b. Heat Exchanger Neural Network Diagram

7-3.   Training a Neural Network

Much of the literature on neural networks describes the mathematical methods that are used to calculate the weights, the number of network layers, and the type of function used for each node. Training neural networks and optimization problems are similar in many ways. After all, the neural network's design is an optimization problem, that is, to search for the best network design to minimize the error between the actual value and the output [7].

Finding those weights is training the network. This is akin to finding the parameter values in a linear regression, but since the neural network model is nonlinear it requires an optimization search [19]. On presenting the data to the network, the training data is forward-propagated through all layers of the network and finally to the outputs. Then, the output is compared to the target value. If the values are significantly different, the weights are adjusted backward through the network. One path of forward and backward propagation through the data set is one training epoch. This process is repeated until the error between the predicted and actual values is satisfactorily small [6].

The most common method for adjusting the weights is the back-propagation algorithm, which updates the weights between the inputs, PEs, and outputs by the error difference between the output of the network and the desired output. This training method assumes that all PEs are responsible for the error. This error is then "propagated backward" through the network, and new weights are calculated based on the direction of steepest descent, or negative gradient of error. Several hidden layers are possible, but in most cases one or two are used. Detailed discussion of this method is available in several of the sources in the references section at the end of this chapter.

The equations illustrating the back propagation method are:

Individual Error

$$E_p = \left(y_p^{pred} - y_p\right)^2 \tag{7-3a}$$

Cumulative Error

$$E = \sum_{p=1}^{all\ data} E_p \tag{7-3b}$$

Gradient

$$\frac{\partial E}{\partial W_{ij}} \tag{7-3c}$$

Gradient Descent Learning

$$W_{ij}^{new} = W_{ij}^{old} - \alpha \frac{\partial E}{\partial W_{ij}} \tag{7-3d}$$

where α defines the step size in the gradient direction.

Training a neural network is considered to be an optimization problem. The back-propagation algorithm is most commonly used to train the network.

Some commercial network training programs use a conjugate gradient descent method in training networks because it improves the learning speed and robustness of the network. This is because, rather than using a fixed step size, the new direction is based on a component of the previous direction [16]. Conjugate gradient descent learning, which is an algorithm for finding the nearest local minimum of a function of n variables, works by constructing a series of line searches across the error surface. It first works out the direction of the steepest descent, in much the same way as back-propagation, and then projects a straight line in that direction, locating a minimum along that line. The directions of the lines, or the conjugate directions, are chosen to ensure that prior directions stay minimized.

One of the best descriptions of these algorithms can be found in T. Masters, *Practical Neural Network Recipes in C++* [21]. Imagine that you are standing at the edge of the Grand Canyon and looking down at the river that flows into the lake a few miles away and to your left. Now suppose that the lake represents the minimum of an error function that you have to find. The most direct route would be diagonally down the side of the cliff. The standard gradient descent method of back-propagation would start you off almost straight down, landing you three-quarters of the way down and a little to the left. The next step will place you halfway up the other side and a few more inches to the left because of the fixed step size. The third step will again take you halfway up the starting side, just slightly more to the left. You will make this side-to-side oscillation until you reach the lake. The conjugate gradient method, on the other hand, would take you directly to the river at the bottom in a few steps of decreasing size. Then it would send you downriver to the lake in a few large steps until the step size decreases as you get close to the lake. Figure 7-3a illustrates the difference between the two methods.

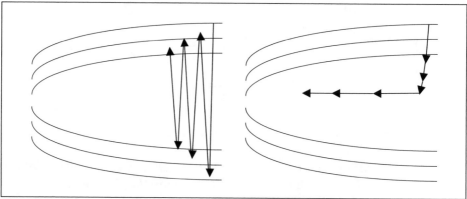

Figure 7-3a. Standard Gradient Descent (Left) vs. Conjugate Gradient (Right)

Other Learning Methods

Back-propagation is a calculation method that is used to select the magnitude of the weights in the network. It is just one of several methods, but the one most frequently described in control literature for network training [7]. Another frequent method to "train" (another way to say regress) a neural network is the Marquardt method [10]. This method makes use of two other methods to optimize data sets: the steepest descent and the Newton method. Both these methods are considered classical mathematical methods for data regression and optimization.

The steepest descent method uses the first-order derivative of the transfer function to determine the direction in which to move. The Newton method uses the second derivative function to calculate the change in weights.

The Marquardt method takes advantage of both these methods. The further away the current solution is from the minima, the faster the steepest descent method will perform. Yet when the weights are closest to the minimum point, the Newton method will approach the solution the fastest. When both these methods for neural network training are used in concert, the Marquardt method will compute the network much faster than will the back-propagation method.

Marquardt is mentioned in many texts on design optimization [7]. We recommend that you become familiar with the Marquardt method if you will be involved in any optimization problems.

7-4. Timing Is Everything

The discussion in the previous section assumed that all the inputs were time independent. In the process applications, however, this is not the case. A process variable at one point in the process may have a strong relationship to the process analytical variable that one wants to model. As an example, a change to a rotary dryer's inlet air temperature will surely affect the dryer product moisture; however, the effect will not be instantaneous. It takes time for the hot gas to travel to the dryer drum, mix with the gasses there, and completely balance the mass transfer of water from the moist product to the air. Neural network programs used for process control should have preprocessing algorithms that can calculate this time delay. This can be done by calculating the cross-correlation coefficients between the inputs and outputs and shifting the time-based data to maximize the correlation.

Developing neural networks with real-time process data requires that the user consider the process's dynamic behavior.

The aspect of process control system neural networks that sets them apart from networks developed using other software is their ability to build a dynamic model. These programs accurately account for the delay between an input change and its impact on the estimated parameter by using a cross-correlation calculation between each upstream measurement selected as a network input and the process output value that is designated as the network output [16]. In order words, since a change in upstream measurement may not be immediately reflected by the process output, the network algorithm delays each input to allow best alignment with the output response. This method is illustrated in figure 7-4a [16]. The time shift, K, which produces the maximum cross-correlation coefficient, is used to build the network.

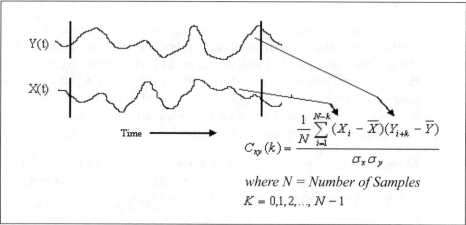

Figure 7-4a. Cross-Correlation Calculations

> *Neural network programs that use cross-correlation*
> *calculations can calculate the input delay times.*

How is this accomplished? The program must calculate the delay for each input. In dynamic processes, there is a timing relationship between a given input and its effect on the output. The program should use the input's historical value rather than its instantaneous value. Identifying this input delay is another optimization problem that involves identifying, for each of the inputs, the delay time associated with that input that has maximum influence on the output.

As a first step in identifying the delays, the cross-correlation between the output and the time-shifted input is calculated for various delays of the input. For an N sample data set, with input and output vectors X and Y, the input vector for delay d is

$$X_d = \left\{ x_d(i), i = 1..N \mid x_d(i) = x(i-d) \right\}$$
(7-4a)

Then the cross-correlation coefficient between the delayed input and output is given by:

$$\rho_d = \frac{Cov(X_d, Y)}{\sigma_{X_d} \sigma_Y}$$

where:

$$Cov(X_d, Y) = \frac{1}{N} \sum_i (x_d(i) - \mu_{X_d})(y_i - \mu_Y) \qquad (7\text{-}4b)$$

where μ and σ denote the mean and standard deviation, respectively. ρ_d is calculated for all possible delays of the input based on process sampling period and final response time (or time to steady state). Figure 7-4b is a plot of the cross-correlation coefficients for various delays of one process input.

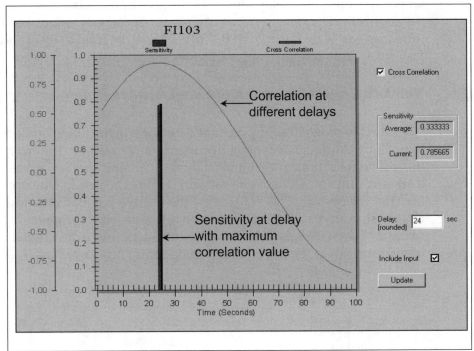

Figure 7-4b. Cross-Correlation Coefficients

The cross-correlation value indicates the magnitude (and sign) of the input's effect on the output. For example, for a simple first-order-plus-dead-time process, most relevant is the historical input value delayed approximately by the *dead time + time constant/2*, since the maximum correlation value occurs at that delay. The second step in delay identification, therefore, is a heuristic-based approach that defines the maximum of the correlation plot. To remove noise and spurious maxima, filtering of the following form may be used:

$$\rho_d = \alpha_1 \rho_d + \alpha_2 (\rho_{d+1} + \rho_{d-1}) \qquad (7\text{-}4c)$$

where:

$$\alpha_1 + \alpha_2 \approx 1; \alpha_1 > \alpha_2$$

In figure 7-4b, the input delayed by 24 seconds has most influence on the output. Once the most significant delay is known, the input data is shifted by that delay to form the data set for training, as shown in equation 7-4a. For example, the pth sample of an n input NN would be of the form:

$$\{x_1(p-d_1), x_2(p-d_2),.., x_n(p-d_n), y(p)\}$$

where d_i denotes the delay for the ith input. Although the delayed input represents the significant process dynamics, secondary delays in a time-series fashion (e.g., $X(d-1)$, $X(d)$, $X(d+1)$) may be used to incorporate the faster-responding higher-order dynamics.

7-5.   Network Generalization: More Isn't Always Better

Consider the plot of training error as a function of the number of epochs in figure 7-5a. Although this error decreases monotonically, the error for a different data set that was not used during training (test set) starts increasing after reaching the minimum. This is because the network just "learns" the training data set and not the underlying mathematical relationship. In neural network lexicon this is called *over training*. The goal of training, however, is to learn to predict the output from unseen real inputs, and not to memorize the training set. This is known as *generalization*.

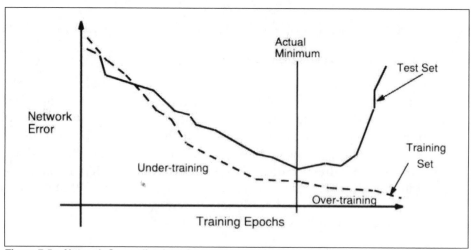

Figure 7-5a. Network Generalization: Cross-Validation with Test Data

To train the network, the user should have three data sets: a training set, a testing set, and a validation set. As discussed earlier, at each epoch the weights are modified based on the error in the training set. In addition, at each epoch, the error on the test set is also calculated to cross-validate the new set of weights. This prevents the network from overfitting the training set. An accepted default split is 80 percent for the training set and 20 percent for the testing set from the available historical data. Once the network has been trained and tested, the third, independent validation set can be used to verify the accuracy of the network predictions on an unseen data set. This may be a standard reference set such as the typical reaction cycle of the bioreactor or data that represents the current operating conditions.

The data set should be separated into three sets: training, testing and validation. The testing and validation sets should not be used for training.

The use of a minimum amount of weights to describe a functional relationship is a principle that applies as equally to neural networks as it does to linear regression techniques. This concept should definitely be considered if the data set is small. Overtraining also results if a large network is used to train, as illustrated in figure 7-5b.

To ensure the best possible model, process neural network programs will train several networks by incrementing the number of hidden neurons. The network that has the minimum cumulative train/test error is then stored for subsequent use. In the case in which two networks yield errors within a tolerance limit, the network that has the fewer number of neurons is given preference.

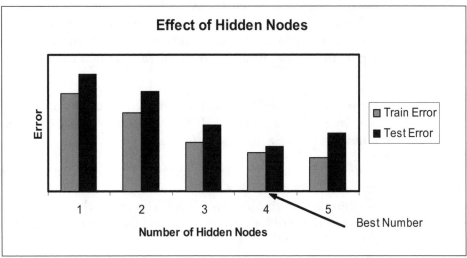

Figure 7-5b. Network Generalization: Prefer Smaller Number of Hidden Nodes

An advanced network development procedure, called *pruning*, starts with a trained network, removes a weight, and then retrains the remaining network, usually with a second data set that is different than the first set. The pruned and retrained network then checks the change in error caused by removing the weight. This sequence is repeated until the error begins to increase, at which time the pruning algorithm stops [11].

Pruning should be performed on all data sets taken from experimental data. For networks that are derived from a computational method, that is, where a neural network is used to describe the results from a complex simulation or calculation, pruning may not reduce the size of the network since all input variables are used to define the network.

7-6.   Network Development: Just How Do You Go about Developing a Network?

The methodology for developing a neural network model can be summarized by the following five steps [15]:

- *Identify the input and output variables.* This is often a difficult step because it is not always obvious what variables will affect the output. It is common in neural network modeling to collect data for all the variables that one believes affect the output and then to use a statistical or neural network–based correlation technique to identify the variables that actually do contribute to the output. The output variable is the value that the network is set up to predict.

- *Collecting data.* Historical data must be collected for each input variable because the model is built from this data. Inaccuracies in time-stamping the lab analytical results is the number one factor for unsuccessful neural networks, and it is impossible to go back and fix existing data sets. The user is strongly advised to set up procedures during data collection that ensure that laboratory results report the proper time at which the sample was drawn from the process.

- *Preprocessing data.* This involves removing outliers and undesired data by using statistical analysis. Without this step, the chances of building a successful model decrease.

- *Training the model.* The data is fed to the network, and the weighted connections between the nodes are adjusted until the model's error is within an acceptable error level. The network architecture, such as the number of hidden neurons and the weights, is determined during this step.

- *Validating the model.* To test the validity of the model, the network is tested using a separate data set to see how well it makes predictions. If the model successfully predicts the output for the verification data set (data that was not used in training) then the model should give correct predictions for the actual system.

Although it is not included as a step, understanding the relationships in the process, and how they are governed, enhances the user's ability to select the correct variables and the data required for training. Ultimately, it is this process understanding that will ensure the robustness of the final application.

Before collecting data, make sure that all required instrumentation works and is not bypassed or configured out of service in the DCS system. If possible, avoid correlated inputs, such as, all tray temperatures in a distillation column. This just adds additional inputs that the training algorithm will have to regress. Use raw uncompressed values without filtering and averaging.

For applications in which an industrial process neural network application is not available, the user will have to manipulate the data in some form as well as perform the training and testing functions. Where an industrial neural network program is available, the process detailed next is frequently automated and a well-designed user interface is developed.

The first step is to organize the data, for example, in an EXCEL® spreadsheet. Then visually remove entries that have empty data variables. Next, check the minimum, maximum, and mean of each data variable. It is important to train the network using the data set that contains the full range of each variable. Since neural networks have nonlinear characteristics, they do not extrapolate well (or at all). The network will not be accurate if the tested variables lie outside the training data envelope.

Do not throw data at a neural network. Ensure data quality by including data that has sufficient variability and covers all operating regions in the training set.

One of the main points of opposition against neural networks is that they dismiss known first-principle methods. This statement is true when considering the network itself, which is a mathematical relationship that would precisely resemble the scientifically derived equations only by accident. However, the user should consider known principles when selecting the inputs to the network. Do not use variables that contain redundant or auto-correlated information. As an example, let's assume that the user is developing a network to model a drying process of a complex solid. Input variables such as air temperature and humidity should be incorporated, as they are logical variables that are known to affect the drying process. Adding wet bulb temperature as an input would serve no useful purpose since the wet bulb temperature has a known relationship to the air temperature and humidity.

> *Consider first-principle-based process knowledge when selecting inputs.*

Once all the data has been checked and measured, separate it into three groups. Use two groups (train and test) if the data set is small compared to the size of the network. Three sets are generally required to validate the network. For best results, the distribution of the training and testing data sets should be close to each other and the entire data set; say, the mean and standard deviation are within 5 percent of each other.

> *It is very important that the data points in test sets be within the values used to train the network. The training set must contain the minimum and maximum points!*

Before starting the training operation, check the population of the data variables. This can be done by plotting a histogram for each input variable over the trained set. If too many data points are the same, the network will learn that one point. This is because the least error squared is weighted heavily at the one point, repeated many times.

Next, each data input and output point are scaled, that is, they are normalized and centered on zero with a standard deviation of one. The uniform scaling, via normalization, of input and output values equalizes the importance of process variables so that learning is not biased to the engineering range. For example, variables like temperature may have a range of a thousand degrees, while others like composition have a range of a few units. Scaling and normalization is done to ensure that each data point is compared to the other data points using the same reference. This is true no matter what algorithm is used to train the network.

A simple rule should be applied when using neural networks. Develop a linear regression first. In many applications, the linear relationship will produce a better correlation than a neural network. One can easily test this by using a simple regression equation of the data and then examining the correlation coefficient.

> *Always consider a simple linear regression before attempting to develop a neural network.*

If the linear relationship does not produce a good fit, then a neural network should be developed. For large data sets that have many inputs, a good start is the back-propagation algorithm. If the data is taken from a simulation or a small data set, then the Marquardt algorithm usually will produce good results quickly.

Another problem that can prevent the network from providing a reasonable relationship is the presence of outliers (data values outside established control limits) in the data set. Training that includes outliers will force the network to a point that does not represent normal operational behavior. One technique for removing outliers is to perform a "reality check" about one-third of the way into the training iterations. This check examines the residuals, the set of individual errors between the training results, and the actual data points. Data points are removed from the data set if they are outside some bound, usually three-and-a-half standard deviations. This is important for large data sets—that is, several thousand points or more—of experimental data. Manually generated data numbers can frequently be entered wrong. Some data points are abnormal for an infinite number of reasons. Within some uses of statistics, the outliers become a major concern. Drug discovery, for example, is very interested in outliers. Outliers are studied more intensively than are the normal cases. For industrial applications there should be no problem with discarding outliers from a training set. This outlier concept can be used for linear regression techniques also.

Once the training is done, check the network against the validation data set. If the network does not appear reasonable and there are an excessive number of weights, consider some of the generalization and pruning techniques, provided there is enough data. If both data quality and quantity issues are addressed, industrial neural network applications generally do a good job of automatically realizing a neural network that can be successfully used for online predictions.

7-7.   Neural Network Example One

A laboratory process uses a proprietary physical property analyzer to analyze the concentration of one of the impurities. For high values of the impurity, the lab used a simple graph to correlate the instrument signal to the concentration. However, at lower concentrations, this relationship did not correlate. Because the process is very complex it was postulated that a neural network could be developed to improve the correlation by combining the process input signals, such as temperature, pressure, level flow rates, and the like, with the analyzer signal. This is an example of a "soft sensor" application.

In this particular example the number of data points was limited. The data was separated into two groups, a training set and a test set. This training data set consisted of 137 entries while the test set consisted of only eight data entries. There were five process data signals plus a constant to make a total of six input points to the network. The output is the impurity concentration in parts per million exiting the process. The test set was *not* part of the training data set in order to preserve the test set validity. The data points were recorded when the process reached steady state. This is very important when the network is used to train a dynamic system; otherwise, historical data should also be included, which can create a large model.

The data was regressed to a single hidden-layer, hyperbolic nodes neural network using the Levenberg-Marquardt method. The data was regressed using an algorithm available over the Internet [9] and written in MATLAB®.

Figure 7-7a shows the test result of the first regression, and figure 7-7b shows the result of the training set. The initial training gave poor results because there were two points at which the impurity analysis was zero. These points were deleted from the data set. Though the results from that set appear to be good—R-squared over 0.9—the network was overtrained, which explains its poor performance in estimating the high concentration. Figure 7-7c is the diagram of the network, consisting of 28 weights, too many weights for such a small data set.

An algorithm provided in the neural network package that "prunes" the network was then used on the first network. This method trims the size of the network and tests it against both the training and test set to obtain the best fit. The results of this pruning session are shown in figures 7-7d, 7-7e, and 7-7f. The network did have difficulty determining the exact concentration at very high levels. This is because there are not many points at this level to train. The final algorithm only uses the nonspecific analyzer signal and one other process signal. The other node is a constant. A way to obtain the confidence of the network is to prepare an EXCEL® spreadsheet that performs the same calculation as the neural network. This way, the user can input several "what-if" scenarios and observe the behavior.

When this information concerning the network was communicated to users, and they became aware of the effect of the other input signal, they recalibrated the analyzer. This required that they not combine data gathered after the calibration change with the old data. When the new data was analyzed, a simple linear regression was used to calculate the impurity concentration. One way in which the old data could have been used with the new data would be to introduce an input to the network: –1

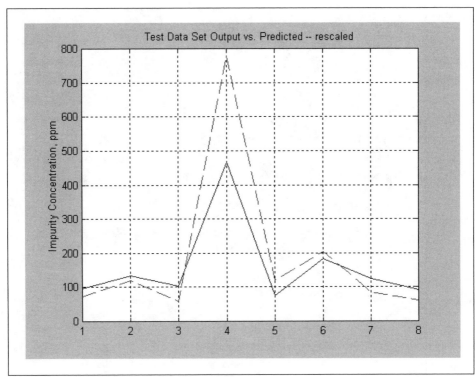

Figure 7-7a. Test Result of the First Regression

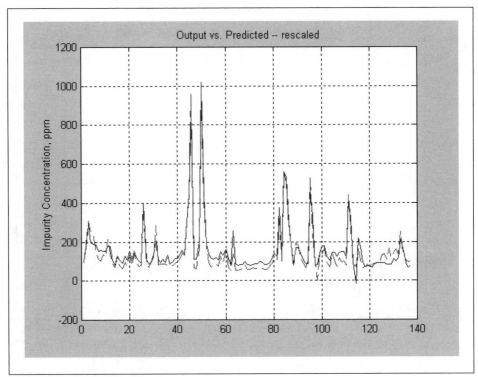

Figure 7-7b. Training Set Results

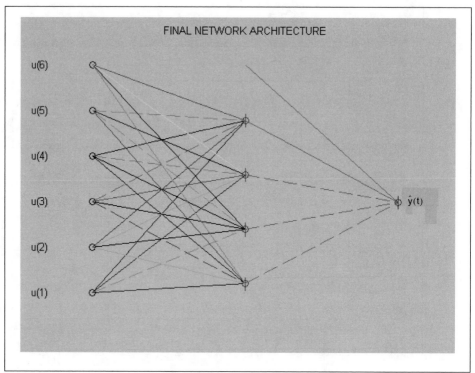

Figure 7-7c. Diagram of the Network

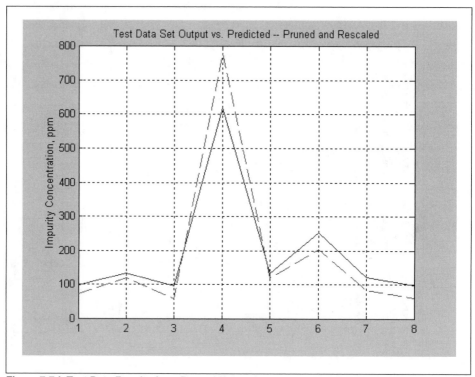

Figure 7-7d. Test Data Results from Pruned Network

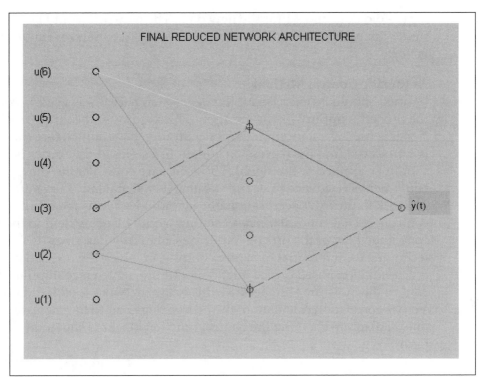

Figure 7-7e. Pruned Network

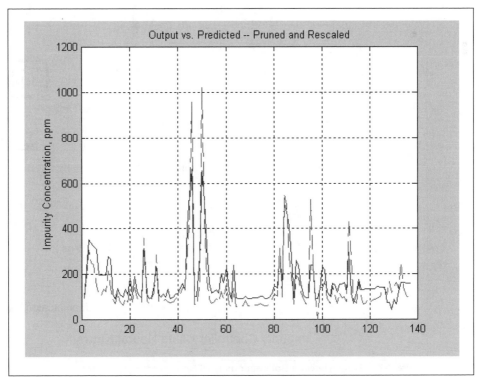

Figure 7-7f. Result of Training Set with the Pruned Network

for the old data and +1 for the new data. The network would be able to train with both sets or add an input that relates an analyzer calibration factor.

Automatic Control Method
Caution should be exercised if this regression network is used to control the process's impurity concentration. This is because the other process signal is used as an input in the concentration equation. Before this signal is introduced into the regression equation, a series of first-order lags should be applied to the signal. These lags are equivalent to the equipment's residence time and the analyzer signal lag. The neural network impurity concentration output would serve as the input to a controller. That controller's reset setting would be equivalent to the total contributed lag of the process input lags. Shinskey describes this technique for inferential moisture controls, in which the implied moisture is a calculation based on the inlet temperature, which is manipulated to control the moisture [12]. Because this neural network toolkit did not have a cross-correlation calculation, the process lags had to be calculated off line. A diagram showing the control function blocks is shown in figure 7-7g.

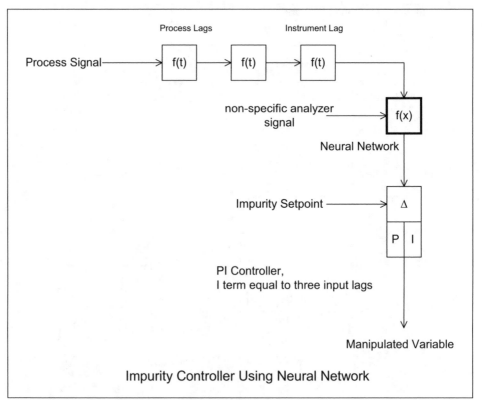

Figure 7-7g. Neural Network Connections to PI Controller

7-8.   Neural Network Example Two

The National Corn to Ethanol Research Center (NCERC) in Edwardsville, Illinois, has a dry grind corn-to-ethanol pilot plant. A free flowing co-product from this process is distiller's dried grains and solubles (DDGS). In addition to stressing the importance of compositional consistency, animal feed nutritionists from academia and industry have indicated that their number one concern about the quality and consistency of DDGS is flowability. The primary issue with DDGS flowability is its propensity to harden when stored or transported by rail cars. Moisture content is of critical importance in bulk solids flowability [18]. Thus, a method to monitor and control the moisture content of DDGS product was needed. To address this need, a neural network for dryer control, which is the portion of the process where moisture content is greatly affected, was developed. In this example, a neural network toolkit for an industrial process control distributed control system (DCS) was used to build the network. This application produced a robust, dynamic model that confirmed that neural networks could be used to model portions of the corn-to-ethanol process.

To begin building a neural network for dryer control, the fifteen variables that were thought to affect the moisture content of the DDGS product were identified and continuously recorded in the DCS data historian at a sampling frequency of once every second. In addition, a sample of the DDGS product was taken every hour and analyzed. The analytically determined moisture content was then entered into the DCS historian. This data and the DCS neural network software allowed heuristic models to be developed for predicting the moisture content of the DDGS product.

A process flow diagram of the drying process at the NCERC is shown in figure 7-8a. As the legend shows, the solid lines correspond to solids flow (grains), the dotted lines are air flow, and the dashed lines are solids recycle, which averaged about 35 percent of the total dryer feed. The DDGS dryer and air heater are industrial, commercially available continuous operating equipment items.

The dryer is a natural gas-fired rotary drum dryer that uses hot air to adiabatically evaporate the water from the DDGS. Within the dryer, baffled walls generate better particle-air contacting by agitating the particles. Wet cake from a decanter centrifuge, with moisture content of 70 percent, is mixed with syrup from an evaporator's bottom discharge. After being mixed with a portion of the final dried DDGS, the mixture, which contains about 40 percent moisture, is dried to a moisture content of 10 percent in the dryer.

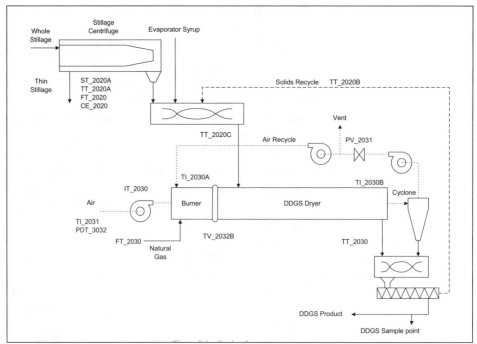

Figure 7-8a. Drying Process

Figure 7-8b. DDGS Dryer

The inlet air is filtered and blown by a combustion air fan to the air heater where it mixes with recycled air. The mixture of wet cake, syrup, and recycled dried product passes through the rotary dryer. Gases from the dryer pass through a cyclone to remove particulates and are recycled or discharged to the thermal oxidizer. The finished DDGS is conveyed by air to storage. The recycling of both exhaust air and product adds complexity to this drying process.

A sample of DDGS product was taken every hour from the point labeled "DDGS Sample Point" in figure 7-8a. The sample was tested for moisture content in a laboratory moisture analyzer and recorded as percentage of moisture. The laboratory instrument works on the thermo-gravimetric principle, which is the method of loss on drying. That is, at the start of the measurement, the moisture analyzer determines the sample's weight, and then the sample is dried at a temperature of 115°C using an internal halogen bulb. When the sample weight is unchanging, the percentage of moisture is calculated using equation 7-8a:

$$\% \; moisture = \frac{wet}{wet + dry} \times 100 \qquad (7\text{-}8a)$$

A temperature profile of TT_2030, the temperature of product exiting the dryer, is given as shown in figure 7-8c. The data used in this plot, which was recorded in the DCS data historian, covers an 80-minute time span. In addition, a profile of DDGS moisture content over the entire nine-day production run is given as figure 7-8d. These plots are included here to show that there is more than a simple temperature function to correlate the outlet moisture. Therefore, this is not a trivial calculation that could be done using a simple linear regression.

Network Construction

Following the steps for building a neural network, the input and output variables were identified. The output variable is the moisture content of the DDGS product. Table 7-8a shows the fifteen variables that were used as inputs, because they were believed to have an affect on DDGS moisture content. These variables are labeled in figure 7-8a.

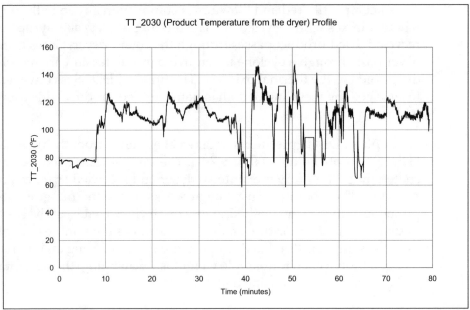

Figure 7-8c. Profile of TT_2030

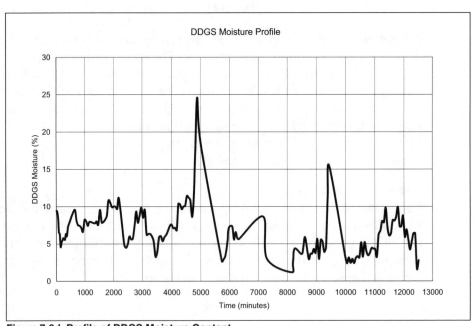

Figure 7-8d. Profile of DDGS Moisture Content

Table 7-8a. Input Variables for the Dryer Control Neural Network

Variable	Description	Units
TT_2020A	Temperature of Whole Stillage Entering the Centrifuge	°F
TT_2020B	Temperature of Distiller's Grains (DG) fed to the DDGS/Feed Mixer	°F
TT_2020C	Temperature of product (in mixer) before being fed to the dryer	°F
FT_2020	Flow out of Centrifuge and Thin Stillage Tank (when operated at SS)	GPM
CE_2020	Percent of maximum centrifuge motor speed	%
ST_2020A	Rotational speed of centrifuge	RPMs
TT_2030	Product/Recycle temperature	°F
TI_2030A	Dryer inlet temperature	°F
TI_2030B	Cyclone temperature	°F
FT_2030	Flow of natural gas into the dryer	SCFH
IT_2030	Current to DDGS/Feed Dryer Combustion Air Fan	Amps
TI_2031	Temperature of air to DDGS/Feed Dryer Combustion Air Fan	°F
PV_2031	Damper position that controls the air recycle rate to the dryer	%
TV_2032B	Set point temperature of dryer	°F
PDT_2032	Pressure on inlet line to DDGS/Feed Dryer Combustion Air Fan	PSIG

We begin our presentation of the results of the four models developed to predict DDGS moisture with the results of the initial sensitivity analysis. A sensitivity analysis identifies the contribution of each input to the estimated parameter. Inputs that do not affect the estimated parameter are eliminated and not used to build the network. DCS neural networks are able to distinguish between a variable having no effect and input data for a variable that covers too narrow a range. In addition, during the sensitivity analysis, the correlation calculation is used to accurately determine the delay between the input change and its affect on the output variable.

The results of the sensitivity analysis performed on all fifteen variables are given in figure 7-8e. This analysis determined the effect that each input variable had on DDGS moisture content and then built the network based on the variables deemed significant. Referring to figure 7-8e, a change in ST_2020A (the centrifuge's rotational speed) takes 8,640 seconds to affect the DDGS product while a change in TT_2020B (the temperature of the distiller's grains fed to the mixer) takes 3,960 seconds. Completing the sensitivity analysis performed on all fifteen variables resulted in the nine-variable network shown in figure 7-8e. The network was trained on these nine input variables:

Nine-Parameter Model
FT_2020 (flow out of the centrifuge),
TT_2020B (temperature of the distiller's grains fed to the mixer),
TI2030B (cyclone temperature),
TT_2020C (temperature of product within the mixer),

ST_2020A (rotational speed of the centrifuge),

TT_2030 (product/recycle temperature from the dryer),

PV_2031 (damper position that controls the air recycle rate to the dryer),

TI_2031 (temperature of air to the dryer fan), and

TI_2030A (temperature of product fed to the dryer)

Of the fifteen input variables, the six that were excluded are:

PDT_2032 (pressure on the inlet line to the dryer fan),

CE_2020 (percentage of the maximum centrifuge motor speed),

IT_2030 (current to dryer fan),

TV_2032B (set point temperature of the dryer),

FT_2030 (flow of natural gas into the dryer), and

TT_2020A (temperature of the whole stillage entering the centrifuge)

The affects of these six parameters were negligible, as denoted by a red "**X**." The table in figure 7-8e lists each variable along with the delay, the sensitivity that DDGS moisture shows toward that variable, and whether it was used in building the network. A delay of –1 is the default for variables that were excluded.

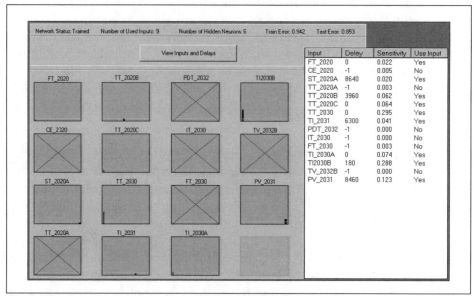

Figure 7-8e. Outcome of Sensitivity Analysis for the Nine-Variable Model

Figure 7-8e also reveals the details of the trained network. There were six hidden neurons, a training error of 0.942, and a test error of 0.853 for this network. Training error is the expected value of the difference between the predictions and the target values within the training data set. Similarly, test error is the expected value of the difference between the predictions and the target values within the test data set. The network architecture is nine inputs, six hidden neurons, and one output (abbreviated 9-6-1). The results of this network are given in figures 7-8f and 7-8g. Figure 7-8f is a plot of the actual and predicted values of DDGS moisture versus sample number, which is a running count of how many DDGS samples were taken. Figure 7-8g is a plot of actual-versus-predicted values of DDGS moisture. From these figures, one can see that the squared error of the predictions was 1.61. Squared error is the expected value of the square of the difference between the estimated and target values. It is essentially used to quantify the variance between the actual and predicted values.

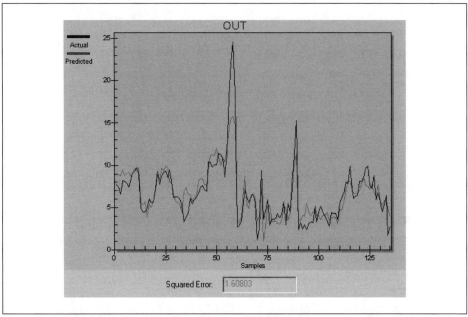

Figure 7-8f. Plot of Actual and Predicted Values of DDGS Moisture-versus-Sample Number for the Nine-Variable Network

After this network was completed, analysis of the results suggested that a better model could be obtained. Considering the small data set (less than 150 sample points in the train/test set and no separate validation set) and the large number of inputs (9), the network that resulted, shown in figure 7-8e, appears to be large and over-specified. Some generalization is therefore required to end up with a smaller and more robust network. The DCS neural network program automatically excluded six variables based on correlation and sensitivity, but in this case further analysis is advised.

Six-Parameter Model

Three additional variables had little effect on DDGS moisture, as indicated by small sensitivity values:

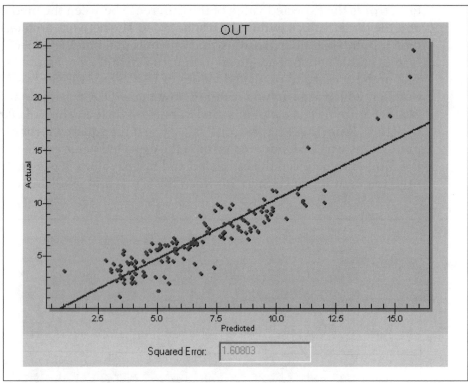

Figure 7-8g. Plot of Actual versus Predicted Values of DDGS Moisture for the Nine-Variable Network

FT_2020 (flow out of the centrifuge),
TT_2020B (temperature of the distiller's grains fed to the mixer), and
ST_2020A (rotational speed of the centrifuge)

These were manually excluded, and this process was repeated until the network was optimized. The ability to select and reject input variables is available to the user with this DCS neural network program. The six-variable network that resulted is shown in figure 7-8h. The architecture that resulted from training based on these six variables was a six-input, four-hidden-neuron, one-output network (6-4-1). The training error was 0.973 and the test error was 0.586. The squared error of the predictions was 1.30.

Four-Parameter Model
Again, utilizing the options available to the user under expert mode in the DCS, two more variables that the network showed little sensitivity toward were excluded: TT_2020C (temperature of the product in the mixer) and TI_2030A (temperature of product fed to the dryer). This resulted in the network shown in figure 7-8k.

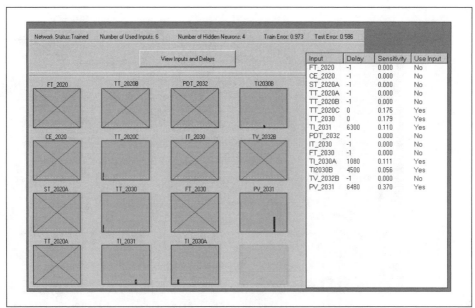

Figure 7-8h. Outcome of Sensitivity Analysis for the Six-Variable Model

The architecture that resulted from training based on the remaining four variables was a four-input, three-hidden-neuron, one-output network (4-3-1). The training error was 0.740 and the test error was 0.694. The results of the network are shown in figures 7-8l and 7-8m. The squared error of the predictions was 1.08.

Figures 7-8n and 7-8o illustrate the method that the DCS neural network program uses to find the delays. These graphs show the cross-correlation calculations for variables PV_2031 (damper position that controls the air recycle rate to the dryer) and TI_2031 (temperature of air fed to the dryer fan). As mentioned, the delay is equivalent to the time at which maximum correlation is achieved.

Two-Parameter Model
Passing air over a moist solid causes the moisture to evaporate, reducing the air temperature. Thus, knowing the temperature of the entering and exiting air might indicate dryer performance. Building a two-parameter model that included only the temperature of air entering and exiting the dryer tested this theory. The results are shown in figure 7-8p.

The architecture that resulted from training based on the dryer inlet and outlet temperature was a two-input, two-hidden-neuron, one-output network (2-2-1). The training error was 1.910 and the test error was 1.578. The results of the network are shown in figures 7-8q and 7-8r. The squared error of the predictions was 3.10.

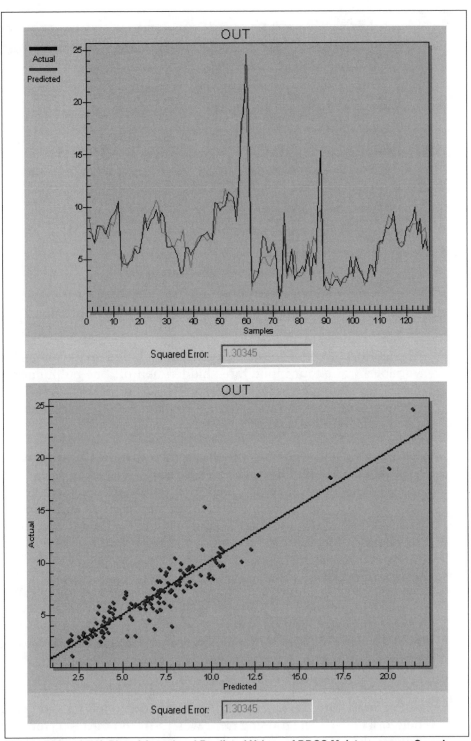

Figures 7-8i and 7-8j. Plot of Actual and Predicted Values of DDGS Moisture-versus-Sample Number for the Six-Variable Network

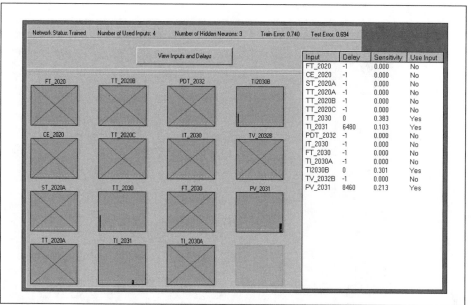

Figure 7-8k. Outcome of Sensitivity Analysis for the Four-Variable Model

Analysis
Table 7-8b gives a detailed summary of the four neural network models developed to predict the moisture content of DDGS.

Table 7-8b. Summary of Dryer Neural Network Results

Input Variables	Hidden Neurons	Training Error	Test Error	Squared Error
9	6	0.924	0.853	1.61
6	4	0.973	0.586	1.30
4	3	0.740	0.694	1.08
2	2	1.910	1.578	3.10

The nine-variable model resulted in a 9-6-1 network. The large number of hidden neurons (six) suggests that the model was overtrained. As the training process begins, the hidden nodes in the network attempt to correlate the major features of the data. As these features have been accounted for, continued training forces the nodes to develop correlations for the secondary features, such as noise, which leads to the large number of hidden neurons. Watching the training and test error during training can prevent this from occurring. This is because overfitting, or overtraining, is evident when the training error decreases despite increased test error [16].

The two-variable model, which was developed to test whether dryer operation could be specified by the inlet and outlet air temperatures,

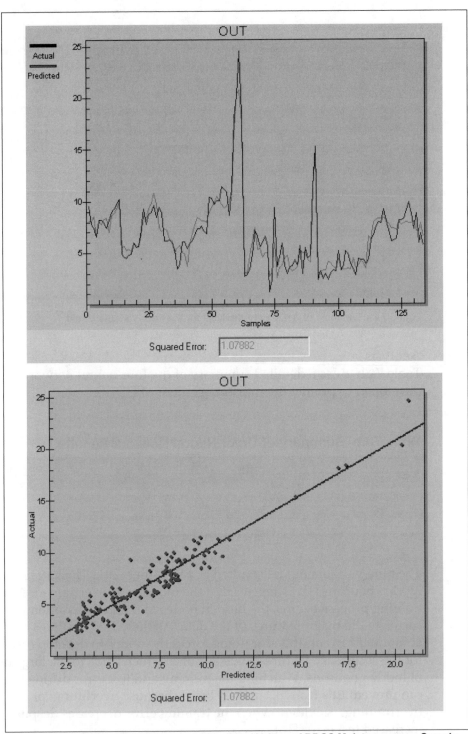

Figures 7-8l and 7-8m. Plot of Actual and Predicted Values of DDGS Moisture-versus-Sample Number for the Four-Variable Network

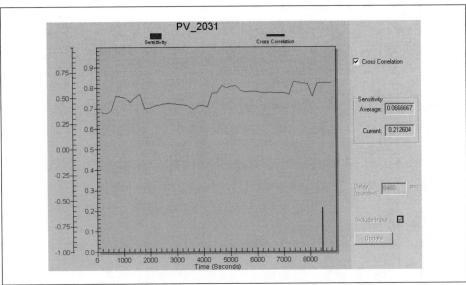

Figure 7-8n. Cross-correlation to Determine Input Delay for Variable PV_2031

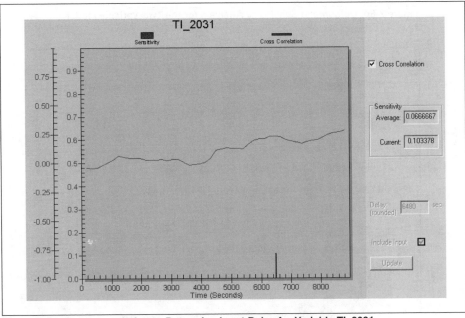

Figure 7-8o. Cross-correlation to Determine Input Delay for Variable TI_2031

failed. This model's poor results are because it lacks a sufficient number of process inputs that impact the output and the resulting small number of neurons. In general, a poor fit is achieved with a smaller number of neurons [16].

The six-variable model yielded accurate predictions. However, reducing the network to the four-variable model gave the best results. Figure 7-8s

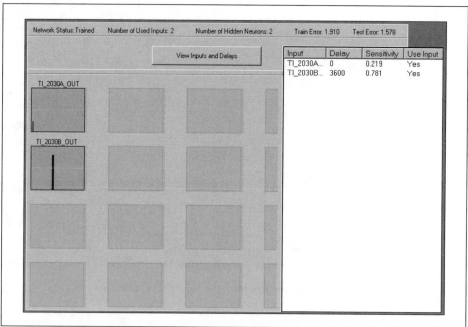

Figure 7-8p. Outcome of Sensitivity Analysis for the Two-Variable Model

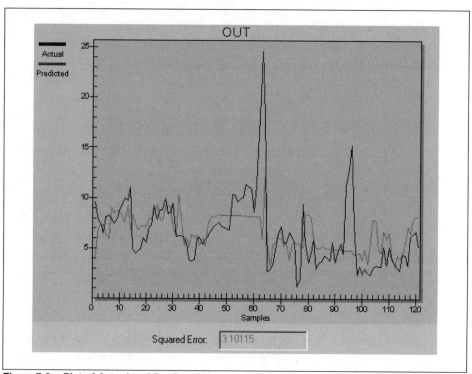

Figure 7-8q. Plot of Actual and Predicted Values of DDGS Moisture versus Sample Number for the Two Variable Network

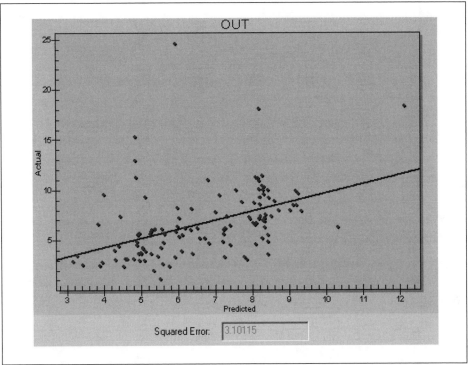

Figure 7-8r. Plot of Actual versus Predicted Values of DDGS Moisture for the Two Variable Network

plots the residuals of the DDGS moisture content, or the difference between the predicted and actual values. The number of occurrences is on the y-axis, and the residuals are on the x-axis. As this figure shows, the residuals follow a normal distribution centered around zero, a perfect prediction, and the predicted values were within 1.3 of the actual values 95 percent of the time.

We can find the model's amount of variance by determining the variability of the moisture analyzer. To accomplish this, a DDGS sample was milled into fine particles, to ensure mixture homogeneity. Ten trial runs were then performed on the mixture using the same temperature profile that was used to test the other DDGS samples. The results are given in table 7-8c.

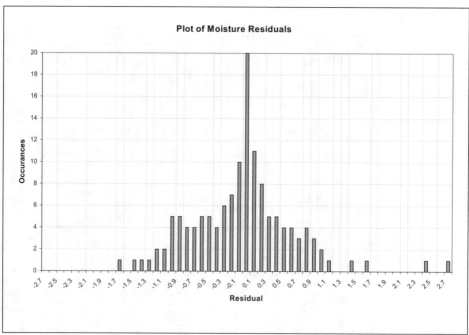

Figure 7-8s. Column Plot of Moisture Residuals for the Four-Variable Model

Table 7-8c. Moisture Analyzer Test Data

RUN	DDGS Moisture
1	10.59
2	10.66
3	10.37
4	10.76
5	11.37
6	10.37
7	10.83
8	10.00
9	10.82
10	10.53

The standard deviation and variance of the moisture analyzer test data were 0.362 and 0.132, respectively. Therefore, the actual variance in the predictions caused by the model was 0.95, which is the difference between the variance reported by the DCS neural network program and the variance in the moisture instrument. If the neural network error was shown to be better than the laboratory analyzer, then the network is overtrained. It is simply modeling the training data set and not the underlying effect.

For "soft sensor" applications, the network variance should be larger than the analyzer variance.

Conclusions

Neural networks are well suited for modeling the drying process at the NCERC. Given the fact that the NCERC process is representative of industry ethanol plants, there is reason to believe that neural networks could be successfully implemented for control purposes in the ethanol industry. In addition, the method of over-specification, then generalization, proved to work well in developing these models. Of the four models developed, the four-input variable model represented the physical system most accurately. It was able to predict the moisture content of DDGS to within 1.3 percent moisture of the actual value 95 percent of the time.

7-9.   Designing Neural Network Control Systems

Another application for neural networks is to use them in some fashion as a process controller, replacing a PID controller. The two methods for control are back-propagation model-based control and inverse model control.

Back-propagation model-based control is a method that incorporates the concept of model predictive control (MPC, also called dynamic matrix control or DMC). In the case of MPC, a linear plant model is used. In applying MPC with neural networks, a nonlinear plant model is used. A series of historical past process variable values are used as the network input, much as with MPC [1].

The concept of inverse neural network control is that if a neural network model can be defined for the process, this model can be inverted and the inverted model can be used for control. This type of controller should be used only when the process is so complex or nonlinear that conventional controls, such as PID, cannot be used. Most of the applications that use inverted networks are for servo metric control [8].

A serious word of caution is in order regarding the use of either of these techniques. Because the neural network uses "trained" parameters and not first-order principles, the network will not give guaranteed performance within any region in which it was not trained. Interlocks should be used to safely shut down the process if any malfunction occurs.

An example of the inverted model technique is shown in figure 7-9a, a conical tank level control.

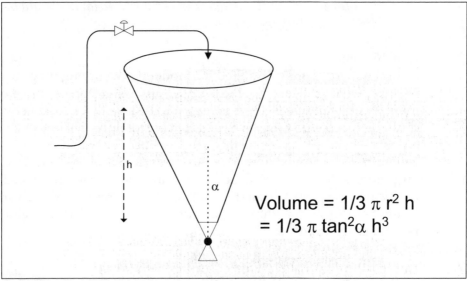

Figure 7-9a. Conical Tank Level

The tank level is a nonlinear process. The outlet flow is a function of the square root of the head while the inlet level change is a function of the inverted square of the head:

$$\frac{dh}{dt} = \frac{1}{\pi \tan^2 \alpha h^2} Q_{in} - \frac{k_{out}}{\pi \tan^2 \alpha h^{1.5}} \tag{7-9a}$$

A neural network was trained to model the cone level and an inverted network was used to control the level. The block diagram and results are shown in figures 7-9b and 7-9c.

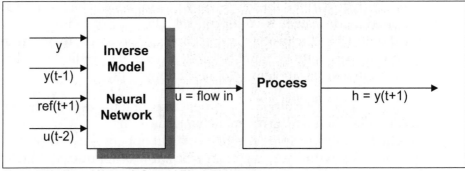

Figure 7-9b. Neural Network Control Block Diagram

The nonlinear behavior can be easily compensated for with this example by simply using a PI controller and cubing the controller's output signal.

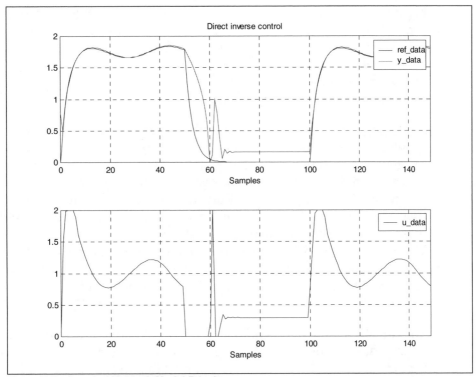

Figure 7-9c. Inverted Network Model Results

The output of this cubing function will have to be divided by a constant to assure that the output will obtain a full-scale value.

For complex process control applications, a more accepted approach, such as a MPC control block, should be considered before implementing a nontraditional method such as inverting a neural network.

7-10. Discussion and Future Direction

Applying neural network technology to different aspects of the process and bioprocess world continues to attract growing interest. This section looks at some of the forward directions being pursued.

Neural Networks for Batch Processes
Most neural network soft sensor applications have been reported in the continuous process world for online *instantaneous* predictions of lab analysis or sampled measurements. However, the process requirements for bioprocess batch applications are often to estimate the final batch properties or the time until end of a batch based on *cumulative* processing conditions during the batch.

Neural networks have been successfully applied to batch processes. One example is developing a neural network to determine a batch end point [25]. For chemical processes, inputs could be reaction kinetics and concentrations, catalyst activities, and physical measurements such as temperature and level. Batch temperature would be important because of the Arrhenius reaction rate equation.

For biological processes, cell growth poses additional complexity. Cell growth is affected by environmental conditions such as pH, dissolved oxygen, and agitation rate. Neural networks can be used in batch processes in the form of "gray boxes" to calculate constants used in first-principle relationships [13]. To build a network to determine the batch end time, you must use historical batch records, where the inputs would be those variables that the users believe effect the reaction or cell growth rate. The network would be trained to learn the time to complete, so the output would be completion time, the first entry in the record would be the completed time for that batch, and the last entry would be zero, that is to say, the end.

The time to steady state would be the longest batch time in the training data set. One could say that the time to steady state for a batch process is the batch reaction time, since it is assumed that the batch is complete at the end of this time period. The user should recognize that this time covers the *reaction* or *growth* period, which may not be the time in any particular phase in the batch sequence.

In applying neural networks to a bioprocess batch reaction, such as the beer fermentation alluded to at the beginning of this chapter, the following characteristics should be considered:

- The rate at which a batch may be processed is a complex function of the initial batch charge, process inputs and conditions during the batch, and in some cases the length of time in which the batch has been processed.

- The completion of batch processing may be calculated by integrating the rate of reaction or rate of processing.

Applying neural net technology to estimate the final batch property or time to complete a batch is, therefore, a good fit because of the nonlinear behavior associated with the rate of reaction or processing. The completion of reaction may be estimated by integrating the rate of reaction (biomass growth, product formation, etc.) over the batch cycle, as shown in figure 7-10a.

One approach to using neural network technology is to estimate the final batch property based on discrete points of time in the batch. This

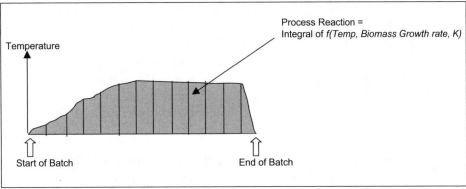

Figure 7-10a. Integrating Bioprocess Reaction Rate

approach has been taken in some implementations. However, even a small application involving only 3 inputs and 50 sampled values spread over the batch produces an over-specified neural network with 150 inputs. It is therefore impractical.

In one innovative method under development, the network input requirements for such bioprocesses are minimized because the neural network is only used to estimate the rate of batch processing. The final batch property can then be estimated by integrating this calculated reaction rate over the batch cycle, as illustrated in figure 7-10b [22].

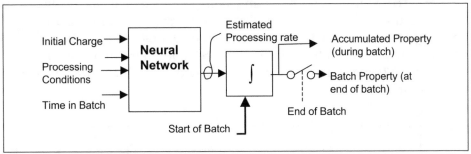

Figure 7-10b. Neural Network Technologies to Predict End of Batch Property

Another promising technology for batch processes is to use *recurrent neural networks*. Unlike the standard feed-forward networks described in this chapter, recurrent networks use feedback from the output as input to the network. Since in batch operations the output depends not only on current and past inputs but also on output values in the past, recurrent networks may be used for end-of-batch predictions.

Neural Networks for PAT Compliance
As mentioned in earlier chapters, PAT or process analytical technology, is a "guidance" issued by the Food and Drug Administration (FDA) (see

section 1-6 in chapter 1) that enables plants to predict and maintain product quality *during* the manufacturing process and not only at the quality assurance (QA) stage. Using neural networks to predict these properties in a continuous fashion without needing to wait for the lab to report analytical results directly assists in PAT compliance. In the long run, the neural network models will also aid in gaining better process understanding. In the near future, neural networks will find increasing PAT applications in bioprocesses.

Correcting Neural Network Predictions Automatically

It is a fact of life that chemical processes change over time. So last year's model will not accurately represent this year's conditions. The black eye that neural network technology has received is the result of operators turning neural networks off because they fail to make reasonable predictions over time. This means that some form of model update is required. It is also established practice that after a soft sensor is deployed, lab measurements and analysis continue as quality checkpoints, even if their frequency is reduced. However, surprisingly, these two facts are rarely combined in a simple way. In fact, several soft sensor manufacturers have proposed online training updates that actually change the fundamentals of the internal network itself. This is not a good idea, both computationally, since training is fairly involved, and because it is impossible to screen the data during online operation. Instead, one simple technique automatically updates the neural network's prediction by comparing it with the time-coincident lab analysis results, as shown in figure 7-10c. This ensures that process drifts and unmeasured disturbances are taken care of, while adaptation does not occur as a result of noise. The automatic bias update should kick off once the cumulative error between network predictions and lab results starts to increase in one direction.

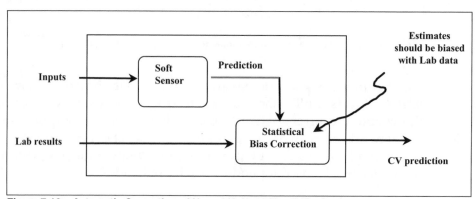

Figure 7-10c. Automatic Correction of Neural Network Predictions

Neural Network Predictions in Closed-loop Advanced Control

The greatest opportunity that neural network technology presents is the ability to use the key quality measures, previously available only infrequently, in full closed-loop (advanced) control and optimization strategies. However, as mentioned earlier, inverting a nonlinear model is fraught with stability issues and is not an advisable control strategy. Recently, industry applications have started using the real-time prediction of the neural network as a control input to a linear MPC. This approach takes advantage of the soft sensor's predictive capability and the multivariable control and optimization strengths of the MPC strategy, combining them into one solution. When they are applied in this fashion, significant reductions in variability—on the order of tens of percent—and huge returns of investment have been reported. For example, in one paper mill the kappa number, a key indicator of paper quality, was being reported only every 1-2 hours from the lab (see figure 7-10d). A neural network soft sensor was first created to provide kappa predictions every minute or so. The output of the neural network was then used in a multivariable MPC strategy to control the kappa number. The mill documented variability reduction of 35 percent in the kappa number [23]. When creating a neural network soft sensor, you should always look ahead to closing the loop on the quality parameters. Some industrial DCS systems make this easy by using a function block approach for the neural network and the MPC development.

Neural Network Extensions for Nonlinear and Transition Control

Recently, significant technical debate has arisen over the use of neural networks for true nonlinear control problems. One of the most promising extensions of the technology is the bounded derivative network (BDN). This has been shown to succeed even during product transitions because the predictive model anticipates lab measurements rather than simply responding to them [24]. The BDN is essentially the analytical integral of a neural network. Since it utilizes a single model and a single set of tuning parameters to cover all operating regimes the required maintenance of such solutions is significantly reduced. Initial applications of BDN to model and control industrial polymer reaction kinetics have shown positive results [24].

7-11. Neural Network Point–Counterpoint

Neural networks are causing controversy in the technical community. Most criticism comes in the form of rejecting inferential models in general where first-principle models have been demonstrated to accomplish the same function. Many of these critics are citing references in the petrochemical industry, where first-principle methods are available. Other

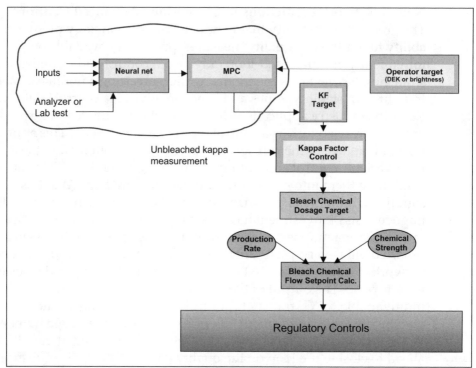

Figure 7-10d. Neural Networks and MPC to Control Kappa Number in a Paper Mill

critics express concern that the processing element takes the form of a sigmoid function that has a "vanishing gain."

First-principle methods are indeed superior in applications where the process is well defined and accurate instruments are installed properly. But, as this chapter has shown, sometimes the instruments are not installed to provide the necessary input to a model. The process's dynamics also must be considered when applying any process model calculation. In these cases, neural networks have been demonstrated to provide excellent inferred indications of analytical process values. Some industries have made better use of neural networks than others. In the interest of fairness, we offer a neural network point–counterpoint:

Point

- A neural network model for NOx was initially developed using all 13 process sensors as inputs. The model was trained and produced a correlation coefficient of 0.98. Excellent agreement between predicted variables and measured variables existed over the entire operating range. A multiple linear regression model was developed using the four most sensitive process sensors as inputs. This predictive model also showed good agreement with the measured NOx emissions.

- Empirical neural network models were applied to a cost-savings problem. The goal was to reduce the amount of natural gas six boilers needed to produce the necessary amount of steam. The neural network model featured sets of manipulated inputs, exogenous inputs, states, and outputs. The manipulated inputs included natural gas and waste gas to feed each boiler, with the objective of minimizing natural gas feed. The exogenous inputs included the status of each boiler (on/off) and the ambient temperature. The states included the steam produced by each boiler. Outputs included total steam produced (constrained to meet the required amount) and total waste gas (constrained not to exceed the daily available limit).

  Once the model was trained and tested on historical data, it could be used to optimize the process to meet the objective. Steam production was optimized for each of the various boilers so that total steam requirements could be met with the least natural gas. For a thirty-day period, the optimization showed that the amount of natural gas could be reduced by 14 percent [27].

- In the words of Sterling, a major gulf coast chemical company, "we are now able to manufacture styrene to much tighter tolerances than was previously possible without adversely impacting operating costs. We believe it is this type of differentiation that will distinguish Sterling from our competitors." Besides improved product quality, Sterling has also increased production capability by 2 percent and decreased energy use by 5 percent annually. This improvement directly contributes potentially millions of dollars to Sterling's top-line growth potential and bottom-line profitability each year.

- In 1990, engineers at Texas Eastman, Longview reported positive results from their use of neural networks to accurately predict how much expensive additive was needed to remove byproduct, thereby reporting savings of 30 percent with no capital cost [14].

Counterpoint

- New EU safety guidelines impact polymer advanced process control (APC) solutions. The "Eunite" task force, assigned by the European Union (EU) to advise on applying "smart" technology in safety critical operations, has recommended against using neural networks in the closed-loop control of manufacturing processes. Their report, "Final Report of the Task Force on Safety Critical Systems," issued in September 2003, describes the dangers in detail in an appendix titled "Product Grade Transitions—Exposing the Inherent and Latent Dangers of Neural Networks in Manufacturing Process Control." The task force concludes: "Difficulties with vanishing and inverting gains are a characteristic of the feed forward neural networks and are better handled by other technologies." Both of these problems have catastrophic implications for a closed-loop controller on a manufacturing facility. The task force also notes: "Closed-loop control of safety-related plant should use transparent models" (i.e., not black boxes such as neural networks). These stark conclusions should be a wake-up call to all process manufacturers currently using or considering using neural networks for closed-loop control of their capital assets. Anticipating that neural networks were not a suitable algorithm for influencing valve movements on manufacturing processes, some companies made a substantial investment in creating a suitable alternative technology that is specifically endorsed by the EU task force on safety critical systems. The full EU report and appendices can be downloaded from the following Web site: www.eunite.org/eunite

- Many process manufacturers invest in inferential "soft sensor" technology, and instead see the advanced control drive the unit further away from optimum operation. One example focusing on the hydrocarbon industry is discussed in the article "How to Lose Money with Inferential Properties" (*Hydrocarbon Processing*, October 2004). "The hydrocarbon industry is remarkably inventive in providing a never-ending catalog of ways to jeopardize the benefits that advanced control should capture," the article notes. "Inferential properties, now commonplace in advanced control projects, offer another rich vein of examples." The article then presents a list of 14 rules that will cause poor performance with a soft sensor [26].

References

7-1.   Bhat, N., and McAvoy, T. J., "Use of Neural Nets for Dynamic Modeling and Control of Chemical Process Systems." *Computer and Chemical Engineering*, I4, no. 4/5 (1990): 573-83.

7-2.   Soucek, B., *Neural and Concurrent Real-Time Systems: The Sixth Generation*. New York: John Wiley & Sons, 1989.

7-3.   Johnson, R. Colin, and Brown, Chappel, *Cognizers, Neural Networks and Machines That Think*. New York: John Wiley & Sons, 1988.

7-4.   Zornetzer, Steven F., *An Introduction to Neural and Electronic Networks*. San Diego: Academic Press, 1990.

7-5.   Hertz, John, Krogh, Anders, and Palmer, Richard G., *Introduction to the Theory of Neural Computation*. Redwood City, CA: Addison-Wesley, 1991.

7-6.   Eberhart, Russell, and Dobbins, Roy W., *Neural Network PC Tools: A Practical Guide*. San Diego: Academic Press, 1990.

7-7.   Reed, Russell D., *Neural Smithing: Supervised Learning in Feedforward Artificial Neural Networks*. Cambridge, MA: The MIT Press, 1998.

7-8.   Nørgaard, M., Ravn, O., Poulsen, N. K., and Hansen, L. K., *Neural Networks for Modelling and Control of Dynamic Systems*. London: Springer Verlag, 2000.

7-9.   MATLAB® Neural Network Toolkit. Technical University of Denmark, Department of Automation, 1997.

7-10.   Marquardt, Donald W., "An Algorithm for Least-Squared Estimation of Nonlinear Parameters." *Journal for the Society of Industrial Math* 11, no. 2, June 1963.

7-11.   Hassibi, B., Stork, D. G., "Second Order Derivatives for Neural Network Pruning: Optimal Brain Surgeon." In *NIPS 5*, edited by S. J. Hansen et al. San Mateo, CA: Morgan Kaufmann, 1993.

7-12.   Shinskey, F. G., *Energy Conservation Through Control*. New York: Academic Press, 1978.

7-13.   Silva, R. G., Cruz, A. J. G., Hokka, C. O., Giordano, R. L. C., and Giordano, R. C., *A Hybrid Feedforward Neural Network Model for the Cephalosporin C Production Process*. Departamento de Engenharia Química, Universidade Federal de São Carlos, São Carlos, Brazil

7-14.   Havener, J. P., Rehbein, D. A., and Maze, S. M., "The Application of Neural Networks in the Process Industry." *ISA Transactions*, 31, no. 4 (1992): 7-13.

7-15.   Ali, M., and Ananthraman, S., "The Emergence of Neural Networks." *Chemical Processing Journal*, September 1995: 30-34.

7-16.   Blevins, Terrence L., McMillan, Gregory K., Wojsznis, Willy, and Brown, Michael, (2003). *Advanced Control Unleashed: Plant Performance Management for Optimum Benefit*. ISA, 2003.

7-17.   Neelakantan R., and Guiver, J. (1998). "Applying Neural Networks." *Hydrocarbon Processing Journal*, September 1998: 91-96.

7-18.   Nokhodchi, A. "An Overview of the Effect of Moisture on Compaction and Compression." *Pharmaceutical Technology Journal*, January 2005: 46-66.

7-19.   Ramachandran, S. and Rhinehart, R., "Do Neural Networks Offer Something for You?" *Engineer's Notebook*, November 1995: 59-64.

7-20.   Kurz, T., Fellner, M., Becker, T., and Delgado, A., "Observation and Control of the Beer Fermentation Using Cognitive Methods." *Journal of the Institute of Brewing* 107, no. 4 (2001): 241-52.

7-21.   Masters, T., *Practical Neural Network Recipes in C++*. San Diego: Academic Press, 1993.

7-22.   Blevins, Terrence L., and Mehta, A., "Apparatus and Method for Batch Property Estimation," U.S. Pat. Pub. No. 2004/0243380, filed May 2003.

7-23.   Pelletier, S., Naiche, F., and Perala, J., "Closing the Critical Control Loops in Bleach Plant with Advanced Process Control Solutions." *PAPTAC Conference Proceedings,* Montreal, Canada, PAPTAC, 2004.

7-24.   Turner, P., Guiver, J., and Lines, B., "Introducing the State Space Bounded Derivative Network for Commercial Transition Control." *Proceedings of the American Control Conference*, Denver, Colorado, 4-6 June, 2003.

7-25.   Slusher, S. M., Bennett, R., and Deitz, D., "Fermentation Application Uses Neural Networks to Predict Batch Time." *ISPE Pharmaceutical Engineering* 14, no. 5, September/October 1994.

7-26.   King, M. J., "How to lose money with inferential properties," Hydrocarbon Processing, October 2004.

7-27.   Proceedings of TAPPI/ISA Conference, 3/01, San Antonio, Texas, USA.

Chapter 8:
Multivariate Statistical
Process Control

Chapter 8

Multivariate Statistical Process Control

Yang Zhang, Research Assistant
Department of Chemical Engineering,
University of Texas, Austin, Texas

Thomas F. Edgar, Professor of Chemical Engineering
Department of Chemical Engineering,
University of Texas, Austin, Texas

8-1. Introduction

In order to meet quality, safety, efficiency, cost, and environmental targets in the process industries, it is often necessary to combine process control and monitoring methods [1]. Process control techniques were discussed in earlier chapters of this book, so this chapter focuses on monitoring techniques. Traditionally, statistical process control (SPC) charts such as Shewhart, CUSUM, and EWMA charts [2] [3] have been used to monitor industrial processes for the purpose of improving product quality. Such techniques are well developed for manufacturing processes and widely used in univariate systems in which a single process variable is monitored. However, industrial process performance usually includes more than one process variable, and univariate SPC charts do not perform as well for these multivariate systems [3-5]. For example, in chemical and biological systems, hundreds of variables can be recorded regularly in a single operating unit, resulting in a large data set that is hard to analyze with univariate methods. Furthermore, because many of the variables may be correlated, they must be considered all together rather than individually. To meet these requirements, multivariate statistical process control (MSPC) methods can be employed to reduce the dimension of the large raw data set while extracting useful information such as the existence of faults or abnormalities.

Since the late 1980s, industry has applied multivariate statistical projection methods to detect abnormal conditions and even diagnose the root cause of abnormal situations [6] [7]. In nearly all applications, two basic techniques are applied: principal component analysis (PCA) and partial least squares (PLS). PCA is a dimensional reduction method that identifies a subset of uncorrelated vectors (principal components) so as to capture most variance in the data. PLS is a decomposition technique that maximizes the covariance between predictor and predicted variables for each component [4]. In some cases, PCA can be applied to a data set before using PLS to reduce the variable dimension for analysis by PLS. This step makes PLS much more efficient to use (Hoskuldsson [8], and MacGregor and Kourti [3] provide more detail on PLS).

MSPC has been applied across many industries during the past ten years, with applications in the areas of semiconductors [9] [10], chemicals [11], mining [12], and petrochemicals. Miletic et al. [13] performed a comprehensive review of the application of multivariate statistics in the areas of steel and pulp and paper. Kourti et al. [14] showed the power of MSPC in polymer processing.

In comparison with these industries, the application of MSPC in the food, biological, and pharmaceutical industry is relatively immature. With online analyzers, plants are able to collect a tremendous amount of historical data. However, efficient tools are needed to manage these data and extract useful information. MSPC is a good candidate for this purpose as well as for achieving the manufacturing goals of high quality and efficiency and low cost [1]. Increasing attention from both academic and industrial researchers has been focused on this promising area. In 2001, Albert and Kinley [15] from Eli Lilly Company applied principal component analysis to an industrial batch tylosin biosynthesis process. In 2001, Lennox et al. [16] published results on different MSPC approaches for a fed-batch fermentation system operated by Biochemie Gmbh. An MSPC application in a pharmaceutical fermentation process was studied by Lopes et al. [17] in 2002. Also, Chiang et al. [18] compared the performance of three MSPC methods in an industrial fermentation process at the San Diego biotech facility of the Dow Chemical Company.

To provide working details on applying MSPC, this chapter is organized as follows. Section 8-2 provides introductory background information on process-monitoring computational algorithms for PCA and a simple example to demonstrate how and why PCA works as a monitoring tool. In section 8-3, the most common PCA approach for a batch process (Multiway PCA) is shown, and some special features of PCA applications to batch processes are discussed. Another type of monitoring method based on nonlinear models, called "model-based PCA," is introduced in section 8-4. Fault detection results for a biological reactor simulation example are given in section 8-5.

Learning Objectives

A. Understand the ideas behind principal component and multivariate statistical process control.

B. Learn how to calculate principal components from a series of correlated data.

C. Recognize different types of PCA and where they should be applied.

D. Know how to choose the number of principal components.

E. Understand the implications of Dynamic Time Warping for variable batch lengths.

F. Learn how to perform different types of unfolding methods and the meaning behind each method.

G. Know which method to select among Multiway PCA, model-based PCA, and super model-based PCA for a practical batch process.

H. Learn how to perform and analyze offline and online Multiway PCA on an industrial batch process.

8-2.   PCA Background

Short History of PCA

In 1873-1874, Beltrami and Jordan independently derived the singular value decomposition (SVD) method that underlies PCA [19]. Meanwhile, the earliest work on PCA was done by Pearson in 1901 and Hotelling in 1933 [19] [20]. However, due to its heavy computational requirements, the spread of PCA was limited until personal computers became available in the 1980s. Since then PCA has been widely used in many areas, including agriculture, biology, chemistry, climatology, demography, ecology, economics, food research, genetics, geology, meteorology, oceanography, psychology, and process control [19]. According to the *ISI Web of Knowledge* (http://isiknowledge.com), nearly 4,000 technical articles used principal component analysis as keywords in 2004 and 2005.

Simple Example Applying PCA

A simple temperature control loop in a bioreactor can be used to illustrate how PCA works. The system consists of a fermenter, a temperature sensor, and cooling water supply. A valve manipulates the flow rate of the cooling water, and the sensor collects temperature data from the fermenter at regular time intervals. Ten pairs of normal condition data and two pairs of known abnormal runs are listed in table 8-2a.

The raw data are plotted in figure 8-2a. Normal operating data are depicted as 'o' while two abnormal data points are depicted as '*'. The first abnormal case represents a mismatch between valve position and temperature while the second case has a higher temperature than normal.

Table 8-2a. Raw data (* = abnormal data).

Valve Position (% max)	Temperature (°C)
10.1	25.0
9.9	25.0
10.0	25.0
9.8	25.3
9.9	25.2
9.8	25.2
10.1	24.8
10.3	24.7
10.0	25.1
10.1	24.7
10.2 *	25.1 *
9.9 *	25.5 *

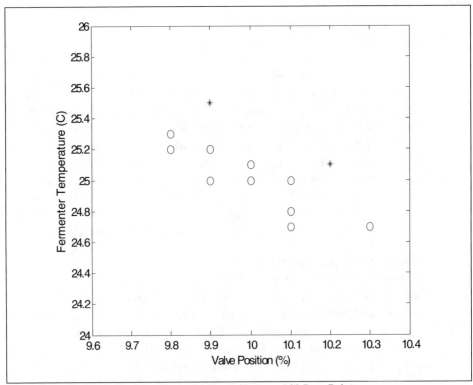

Figure 8-2a. Raw Data Showing Normal (o) and Abnormal (*) Data Points

Control charts of valve position and temperature are shown in figures 8-2b and 8-2c. The upper control limit (UCL) and lower control limit (LCL) are calculated by

$$X_{UCL,LCL} = \overline{X} \pm 2\sigma$$

($\overline{X}$ is the mean and s is the standard deviation of the data), which indicate bounds for normal operation. It can be seen that all valve position data fluctuate around the mean value, while temperature values also appear to be normal ("in control" using SPC terminology) except the last data point. As a result, the first abnormal reading cannot be detected by either control chart.

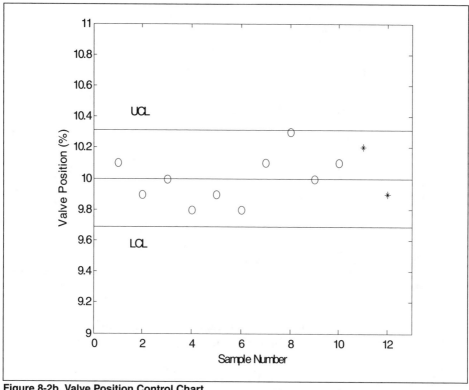

Figure 8-2b. Valve Position Control Chart
Note: The middle line is the average value while the upper line is the UCL (upper control limit), the lower line is the LCL (lower control limit). 'o' represents normal conditions and '*' indicates abnormal cases. All data lie within the UCL and LCL.

In industrial applications, single-variable control charts such as figures 8-2b and 8-2c are frequently used. However, this simple example clearly illustrates that such charts have a deficiency when monitoring correlated variables. Therefore, independent variables are preferred instead of the original variables. PCA provides a way to project the raw data to the principal component subspace and obtain a series of uncorrelated variables called principal components (PCs). A detailed procedure is presented below. Let us now walk through step-by-step sample calculations.

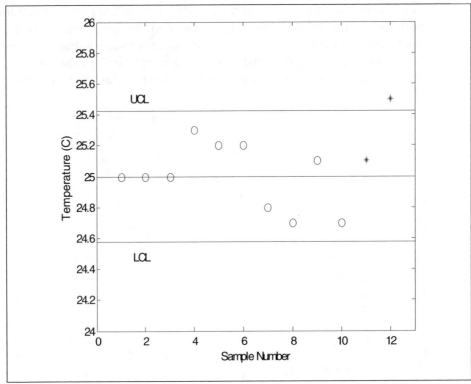

Figure 8-2c. Temperature Value Control Chart
Note: 'o' represents normal conditions and '*' indicates abnormal cases. Sample 12 exceeds the UCL.

To carry out PCA on the raw data, first we calculate the mean, variance, and covariance ($\mathbf{S}$) of the raw data ($\mathbf{X_{raw}}$) after subtracting the mean:

$$\overline{\mathbf{X}}_{raw} = \begin{bmatrix} \overline{\mathbf{X}}_{1,raw} \\ \overline{\mathbf{X}}_{2,raw} \end{bmatrix} = \begin{bmatrix} 10.0 \\ 25.0 \end{bmatrix} \tag{8-2a}$$

The deviations from the means of $\mathbf{X_1}$ and $\mathbf{X_2}$ are as follows:

$$X_{1i} = X_{1i,raw} - \overline{X}_{1,raw} \quad i = 1,2,...,m$$

$$X_{2i} = X_{2i,raw} - \overline{X}_{2,raw} \quad i = 1,2,...,m$$

The resulting data matrix is then denoted as $\mathbf{X}$ ($\mathbf{X_1}$ = valve position deviation and $\mathbf{X_2}$ = temperature deviation), which has zero mean:

$$\mathbf{X} = \begin{bmatrix} X_{11} & X_{12} & ... & X_{1m} \\ X_{21} & X_{22} & ... & X_{2m} \end{bmatrix} \tag{8-2b}$$

where m is the number of observations. Row 1 is the mean-centered valve position data, and row 2 contains the mean-centered temperature data. An important property of $\mathbf{X}$ is its covariance:

$$\mathbf{S} = \mathrm{cov}(\mathbf{X}) = \frac{\mathbf{X}^T\mathbf{X}}{m-1} = \begin{bmatrix} 0.0244 & -0.0289 \\ -0.0289 & 0.0444 \end{bmatrix} \tag{8-2c}$$

The covariance matrix is important since the off-diagonal terms represent the relationship between each variable (Note: $\mathbf{S}$ is symmetric). In this simple case, $s_{12} = -0.0289$, and the correlation coefficient is

$$\rho_{12} = \frac{s_{12}}{(s_{11}s_{22})^{0.5}} = -0.88$$

which indicates that there is a strong linear relationship between valve position and temperature because its absolute value is close to 1.0.

Mean centering is important when calculating the covariance matrix; hence the covariance matrix and PCA derive from the data variation rather than from the absolute data values.

In process monitoring, independent variables are preferred over correlated ones. Thus, we transform the correlated data to a new subspace so that the off-diagonal terms of the new covariance matrix are zero (no correlation between the new variables). Orthogonal transformation provides a way for $\mathbf{S}$ to be converted to a diagonal matrix (Λ) such that the off-diagonal terms are zero [7]:

$$\mathbf{U}^T\mathbf{S}\mathbf{U} = \Lambda \tag{8-2d}$$

$$\mathbf{U}^T\mathbf{U} = \mathbf{I} \tag{8-2e}$$

where $\mathbf{I}$ is the identity matrix and $\mathbf{U}$ is called the eigenvector matrix. The rows and columns of $\mathbf{U}$ are orthonormal, so this transformation effectively forms two new variables that are linear combinations of the original variables x_1 and x_2. The diagonal terms of Λ are the eigenvalues, and the columns of $\mathbf{U}$ are the eigenvectors of the covariance matrix. Using available mathematical software such as MATLAB [21] (command: [u, lambda] = eig(S)), Λ and $\mathbf{U}$ can be calculated easily.

For S in equation 8-2c, $\mathbf{U} = \begin{bmatrix} -0.58 & -0.81 \\ 0.81 & -0.58 \end{bmatrix}$, $\Lambda = \begin{bmatrix} 0.065 & 0 \\ 0 & 0.0039 \end{bmatrix}$ and:

$$\mathbf{U^TSU} = \Lambda = \begin{bmatrix} -0.58 & 0.81 \\ -0.81 & -0.58 \end{bmatrix} \begin{bmatrix} 0.0244 & -0.0289 \\ -0.0289 & 0.0444 \end{bmatrix} \begin{bmatrix} -0.58 & -0.81 \\ 0.81 & -0.58 \end{bmatrix} = \begin{bmatrix} 0.065 & 0 \\ 0 & 0.0039 \end{bmatrix} \qquad (8\text{-}2f)$$

At this stage, principal axis transformation can be applied to transform the correlated raw data into independent values, which can be described as follows:

$$\mathbf{Y} = \mathbf{U^T}(\mathbf{X_{raw}} - \overline{\mathbf{X}}_{raw}) \qquad (8\text{-}2g)$$

The matrix of eigenvectors $\mathbf{U}$ is called the "projection matrix," and the transformed variables $\mathbf{Y}$ are called "principal components" (PCs). Take the first pair of measurements as an example:

$$\mathbf{Y} = \begin{bmatrix} Y_1 \\ Y_2 \end{bmatrix} = \mathbf{U^T}(\mathbf{X_{raw}} - \overline{\mathbf{X}}_{raw}) = \begin{bmatrix} -0.58 & 0.81 \\ -0.81 & -0.58 \end{bmatrix} \left(\begin{bmatrix} 10.1 \\ 25.0 \end{bmatrix} - \begin{bmatrix} 10.0 \\ 25.0 \end{bmatrix} \right) = \begin{bmatrix} -0.058 \\ -0.082 \end{bmatrix}$$

Λ gives the variance of transformed variables, and these two PCs are linearly independent (covariance equal to zero). To make the two PCs have the same scale or length, another way of defining PCs can be used:

$$\mathbf{Z_i} = \frac{\mathbf{U^T}(\mathbf{X_{raw}} - \overline{\mathbf{X}}_{raw})}{\lambda_{ii}^{0.5}} \qquad (8\text{-}2h)$$

By dividing the normalization factor based on the diagonal elements of Λ (λ_{ii}), the new PCs will have a unit variance. This is quite useful because the new artificial variables (PCs) are not only independent but have unit variance as well. The values of the normalized PCs are listed in table 8-2b, and corresponding control charts of the transformed dataset $\mathbf{Z_i}$ are plotted in figures 8-2d and 8-2e. Both abnormal operation conditions are clearly identified in figure 8-2e. In both figures, since the standard deviation of each variable is one, the UCL and LCL are 2 and –2, respectively, ($Z_{UCL,LCL} = \overline{Z} \pm 2\sigma = 0 \pm 2$).

Table 8-2b. Projected Data (Z_1 and Z_2).

PC 1 (Z_1)	PC2 (Z_2)
-0.2275	1.3088
0.2275	1.3088
0.0000	0.0000
1.4134	-0.1782
0.8664	-0.5551
1.0939	0.7537
-0.8664	0.5551
-1.6409	-1.1306
0.3195	-0.9319
-1.1859	1.4870
-0.1355*	-3.5495*
1.8248*	-3.3509*

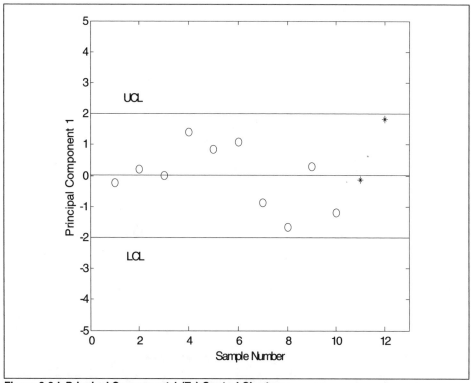

Figure 8-2d. Principal Component 1 (Z_1) Control Chart.
Note: 'o' represents normal conditions, and '*' indicates the abnormal cases. All data lie between the UCL and LCL.

Both abnormal points follow the trend of the process changes (see figure 8-2a and table 8-2b), thus figure 8-2d indicates that the process is in control. However, the correlation between the two variables for the first abnormal case (valve position: 10.2; temperature: 25.1) differs from the normal behavior (i.e., the valve opens and temperature drops, which indicates a negative process gain). For that reason, figure 8-2e shows there is a fault in the system. For the second abnormal case (valve position: 9.9; temperature: 25.5), the temperature should be higher but closer to 25. The temperature is clearly higher than 25 but not close to 25. In other words, the trend of changes between temperature and valve position is followed but not exactly. Note that in figure 8-2d, point 12 is close to the UCL. At the same time, the temperature change relative to the valve position is so large that it exceeds the control limit of figure 8-2e. The role of the projection matrix **U** will be further illustrated in our analysis in *PCA algorithm* section below.

In the foregoing analysis, the covariance matrix (**S**) is used to explicitly show whether the variables are correlated or not. However, in industrial analysis, projection can be carried out directly without explicitly

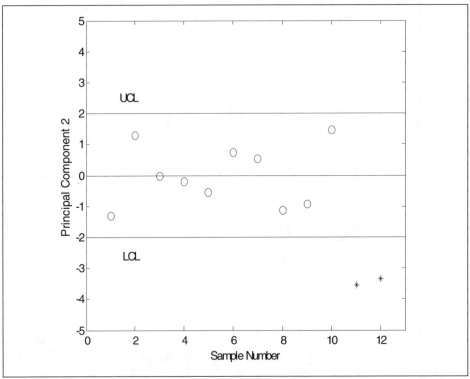

Figure 8-2e. Principal Component 2 (Z_2) Control Chart.
Note: 'o' represents normal conditions, and '*' indicates the abnormal cases. The last two data points are identified as abnormal on this control chart.

computing the covariance matrix. In another words, **X** can be decomposed into the product of two matrices by:

$$\mathbf{X} = \mathbf{TP}^{\mathrm{T}} \qquad\qquad (8\text{-}2\mathrm{i})$$

T and **P** are called "score" and "loading" matrices and have dimensions ($m \times r$) and ($n \times r$), respectively. Modifying equation 8-2d, we obtain:

$$\mathbf{S} = \frac{\mathbf{X}^{\mathrm{T}}\mathbf{X}}{m-1} = \mathbf{U\Lambda U}^{\mathrm{T}} \qquad\qquad (8\text{-}2\mathrm{j})$$

in which m is the number of observations. Substituting equation 8-2i into equation 8-2j, we obtain:

$$\mathbf{S} = \frac{\mathbf{X}^{\mathrm{T}}\mathbf{X}}{m-1} = \frac{\mathbf{PT}^{\mathrm{T}}\mathbf{TP}^{\mathrm{T}}}{m-1} = \mathbf{U\Lambda U}^{\mathrm{T}} \qquad\qquad (8\text{-}2\mathrm{k})$$

Define $\dfrac{\mathbf{T}^{\mathrm{T}}\mathbf{T}}{m-1} = \mathbf{\Lambda}$

and, thus, the loading matrix (P) is the same as the projection matrix (U). The score matrix (T) can be calculated by postmultiplying equation 8-2i by P (using $\mathbf{P^T P} = \mathbf{I}$, where P is the eigenvector matrix of the covariance matrix):

$$\mathbf{T} = \mathbf{X} \cdot \mathbf{P} \tag{8-2l}$$

For this simple example, the first component of the score matrix is as follows:

$$t_1 = \mathbf{X_1} \cdot \mathbf{P} = [(10.1 - 10.0), (25.0 - 25.0)] \begin{bmatrix} -0.58 & -0.81 \\ 0.81 & -0.58 \end{bmatrix} = [-0.058, -0.0815]$$

In this way, the score of all samples can be calculated. The loadings p_j contain the information on relationships between different rows, while the scores t_j describe the relationship between columns. We will discuss how score and loading matrices are used and present a more efficient calculation method below.

The control charts of independent variables (PCs) work fairly well for this simple example. However, in a real industrial application, a process unit might have hundreds of variables, so many control charts with a "false positive" error may result.

> *A "false positive" error describes the situation in which the system data do not contain a fault but control charts indicate there is one.*

As a result of the "false positive" error problem, squared prediction error (*SPE*) and Hotelling's T^2 control charts are used to give a more definitive answer for a multivariate process. The detailed definition and calculation of *SPE* and Hotelling's T^2 for these multivariate problems are given later.

Next, we develop the values of the Hotelling's T^2 control chart. Returning to the simple example, the points in figure 8-2f are calculated by:

$$T_j^2 = Z_j^T \cdot Z_j = Z_{1j}^2 + Z_{2j}^2 \tag{8-2m}$$

T_j^2 is positive for all j and thus has a minimum value of zero. The upper control limit of T^2 is related to the F distribution and can be calculated by:

$$T_\alpha^2 = \frac{k(m-1)}{m-k} F_{k,m-k;\alpha} = 10.03 \text{ (T2 UCL)} \tag{8-2n}$$

k is the number of PCs, m is the number of normal condition data points, and α is the percentage confidence limits expressed as a fraction $(0<\alpha<1)$. In this example, $k = 2$, $m = 10$, and $\alpha = 0.95$, indicating that there are two PCs, ten data pairs, and 95 percent confidence limits.

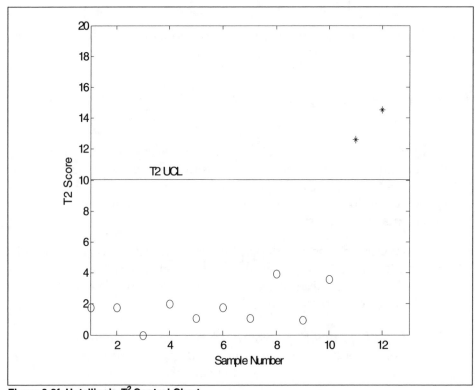

Figure 8-2f. Hotelling's T^2 Control Chart
Note: 'o' represents normal conditions and '*' indicates abnormal case.

The *SPE* control chart focuses on the magnitude of the sample projection on the residual space. For this simple case, the number of PCs is two, which is the same as original dataset, so there are no remaining vectors for correlation because we started with only two variables; hence, the residual is the null space. For more complex problems there can be many original variables, and we might only select fewer than three principal components. Therefore, we will give an example illustrating the use of the *SPE* control chart in section 8-5, in the bioreactor example, which has a large number of process variables.

To provide a geometric interpretation of principal component projection, a control ellipse is constructed for the original variables x_1 and x_2. Combining equations 8-2h and 8-2m, we get:

$$T^2 = (\mathbf{X}_{raw} - \overline{\mathbf{X}}_{raw})^T \frac{\mathbf{U}}{\mathbf{\Lambda}^{0.5}} \frac{\mathbf{U}^T}{\mathbf{\Lambda}^{0.5}} (\mathbf{X}_{raw} - \overline{\mathbf{X}}_{raw}) = (\mathbf{X}_{raw} - \overline{\mathbf{X}}_{raw})^T \frac{(\mathbf{U}^T\mathbf{U}) \cdot (\mathbf{U}^T\mathbf{U})}{\mathbf{U}^T\mathbf{\Lambda}\mathbf{U}} (\mathbf{X}_{raw} - \overline{\mathbf{X}}_{raw}) \quad (8\text{-}2o)$$

$$= (\mathbf{X}_{raw} - \overline{\mathbf{X}}_{raw})^T \mathbf{S}^{-1} (\mathbf{X}_{raw} - \overline{\mathbf{X}}_{raw})$$

It should be noted that $\mathbf{S}^T = (\mathbf{U}^T\mathbf{\Lambda}\mathbf{U})^T = \mathbf{U}\mathbf{\Lambda}^T\mathbf{U}^T = \mathbf{U}\mathbf{\Lambda}\,\mathbf{U}^T = \mathbf{S}$ and $\mathbf{U}^T\mathbf{U} = \mathbf{I}$.

This leads to:

$$\frac{s_1^2 s_2^2}{s_1^2 s_2^2 - s_{12}^2} \cdot \left[\frac{(x_1 - \bar{x}_1)^2}{s_1^2} + \frac{(x_2 - \bar{x}_2)^2}{s_2^2} - \frac{2 s_{12}(x_1 - \bar{x}_1)(x_2 - \bar{x}_2)}{s_1^2 s_2^2} \right] = T_{k,m,\alpha}^2 \qquad (8\text{-}2p)$$

For our example, it should be:

$$\frac{0.0244 \times 0.0444}{0.0244 \times 0.0444 - (-0.0289)^2} \left[\frac{(x_1 - 10)^2}{0.0244} + \frac{(x_2 - 25)^2}{0.0444} - \frac{2(-0.0289)(x_1 - 10)(x_2 - 25)}{0.0244 \times 0.0444} \right] = 10.03$$

which simplifies to:

$$17.83(x_1 - 10)^2 + 9.80(x_2 - 25)^2 + 23.22(x_1 - 10)(x_2 - 25) - 1 = 0$$

This ellipse is depicted in figure 8-2g. The slopes of the major and minor axes are as follows:

$$m_{major} = \frac{\mathbf{U}(2,1)}{\mathbf{U}(1,1)} = -1.4044$$

$$m_{minor} = \frac{\mathbf{U}(2,2)}{\mathbf{U}(1,2)} = 0.7121$$

The lengths of the semi-major and semi-minor axes are:

$$l_{major} = \sqrt{\mathbf{\Lambda}_{11} T_{upper}^2} = \sqrt{(0.065)(10.03)} = 0.8076$$

$$l_{minor} = \sqrt{\mathbf{\Lambda}_{22} T_{upper}^2} = \sqrt{(0.0039)(10.03)} = 0.1971$$

The center of the control ellipse is at: $x_1 = 10.0$; $x_2 = 25.0$ (the mean values of $\mathbf{X_1}$ and $\mathbf{X_2}$).

The two abnormal points lie outside of the ellipse.

For this two-variable example, the control ellipse not only shows whether the process is in control but also provides a geometric figure to explain how principal projection works. However, for high-dimensional systems, the control ellipse cannot be displayed; hence, SPE and T^2 charts are typically used in process quality monitoring.

Besides the issues mentioned in this example, other important topics in PCA include, among others: How to choose the number of PCs? How to carry out PCA on continuous and batch processes? What do *SPE* and T^2 indicate? These topics will be discussed in the remaining sections of this chapter.

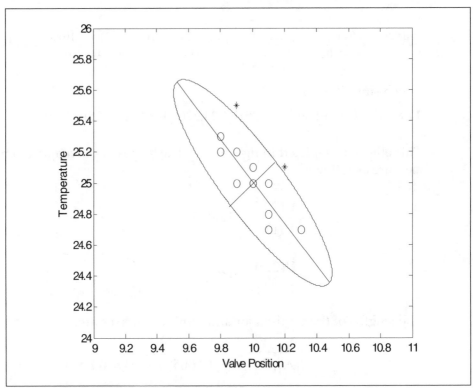

Figure 8-2g. Control Ellipse for Simple Example.
Note: 'o' represents normal conditions and '*' indicates abnormal case.

PCA Algorithm
Because the statistical background of PCA is described in detail by Jolliffe [19] and Jackson [20], this section will mainly focus on the algorithm and the steps for applying PCA for both online and offline analysis.

1. Preprocessing Data

For continuous processes, $\mathbf{X_{raw}}$ is an $m{\times}n$ data matrix whose columns correspond to the process variables (sensors) and whose rows correspond to the samples at each time. Then, $\mathbf{X_{raw}}$ is mean-centered by subtracting the mean value of each column. It is then rescaled by dividing the standard deviation before performing PCA. The data matrix after preprocessing is denoted as $\mathbf{X}$. Centering is performed so the PCA model represents the deviation from the mean, and the scaling step is necessary so each variable has equal weight in further analysis. If different weights are needed for some variables, then a diagonal weighting function ($\mathbf{W}$) multiplies the raw data after mean centering [22].

2. Decomposing Data

After preprocessing, $\mathbf{X}$ can be decomposed into

$$\mathbf{X} = \mathbf{TP^T} + \mathbf{E} \tag{8-2q}$$

using either singular value decomposition (SVD) or the nonlinear iterative partial least squares (NIPALS) algorithm. $\mathbf{T}$ and $\mathbf{P}$ are the score matrix ($m{\times}r$) and loading matrix ($n{\times}r$) ($r = min(m,n)$), respectively. $\mathbf{E}$ is the residual matrix that results when fewer than r of the principal components are used in the model. SVD is an efficient method for finding principal components (PCs). However, when the dimension of $\mathbf{X}$ is large (rank($\mathbf{X}$)>1000), SVD can be time consuming. NIPALS is more efficient since the PCs are calculated sequentially. The theory and algorithm of SVD can be found in mathematics and statistical modeling books (e.g., [19] [23]). There is an svd.m file in MATLAB [21] that provides the source code for performing SVD. In the biological fermentation system discussed in section 8.5 of this chapter, the data matrix is very large. Thus, NIPALS is the best approach; we will describe its algorithm next [22] [24]:

1. Choose any column x_i from X and set it equal to t_j as the initial value.

2. $p_j^T = \dfrac{t_j^T \mathbf{X}}{t_j^T t_j}$

3. $p_{jnew}^T = \dfrac{p_{jold}^T}{\| p_{jold}^T \|}$

4. $t_j = \dfrac{\mathbf{X} p_j}{p_j^T p_j}$

5. If t_j converges, go to step 6; otherwise, go to step 2.

6. $\mathbf{X} = \mathbf{X} - t \cdot p^T$

7.   Go to step 1 to calculate the next PC.

Note that in step 6, the data residual is calculated based on using the current principal component. The calculated p_j vectors correspond to the eigenvectors of the covariance matrix of $\mathbf{X}$, which is defined as follows:

$$\mathbf{S} = \text{cov}(\mathbf{X}) = \frac{\mathbf{X}^T\mathbf{X}}{m-1} \tag{8-2r}$$

So, for each p_j:

$$\text{cov}(\mathbf{X})p_j = \lambda_j p_j \tag{8-2s}$$

where λ_j is the j$^\text{th}$ eigenvalue of the covariance matrix of X and the solution of NIPALS is in descending order ($\lambda_1 > \lambda_2 > ... > \lambda_r$). As we pointed out earlier, the loadings p_j contain the information on relationships between variables, while the scores t_j describe the relationship between samples. The magnitude of λ_j describes the amount of information captured by the corresponding t_j and p_j. Therefore, the set of t_1 and p_1 (corresponding to λ_1) contains the most variation in $\mathbf{X}$. In general, the first few pairs of t_j and p_j capture most of the variance.

The $p_j's$ are orthonormal:

$$p_i^T p_j = \begin{Bmatrix} 0, i \neq j \\ 1, i = j \end{Bmatrix}$$

And the $t_j's$ are orthogonal, namely:

$$t_i^T t_j = 0, i \neq j \cdot t_i^T t_j = \lambda_i, i = j$$

Furthermore, by NIPALS the magnitude of λ_j is already in descending order.

3. Choosing the number of principal components
The first k pairs of loadings and scores are stored in matrices T and P, and these pairs form an optimal model of the system. Therefore, choosing the minimum number of PCs (k) is an important decision. The key idea in this step is to select the number of PCs that will accurately reflect the system behavior and also filter process noise. Typically, for a complex industrial system, more than one method is used and compared to reach a final conclusion. For example, in [25], a simulation example was analyzed by different methods; the results are given in table 8-2c. It can be seen that different methods reach different conclusions regarding the number of

principal components needed for three different unit operations. Furthermore, some methods show ambiguous results or no solution at all. Finally, it is worth noting that usually two or three PCs are sufficient to explain most of the variance in the system.

> *The choice of how many PCs to use is somewhat subjective and depends on the user's experience with the process. Some selection methods—for example, VRE (variation of reconstruction error), parallel analysis and cross-validation— may show better reliability and effectiveness.*

4. Projecting Data

Based on the information thus far, a sample x ($n \times 1$) is projected to the principal component space ($\hat{x}$) and residual space ($\tilde{x}$) by:

$$\hat{x} = \mathbf{PP}^{\mathbf{T}} x \tag{8-2t}$$

$$\tilde{x} = (\mathbf{I} - \mathbf{PP}^{\mathbf{T}}) x \tag{8-2u}$$

Table 8-2c. Number of PCs Determined from Each Analyzed Method (table 1 in [25])

Method	Reactor	Boiler	Incinerator
Cumulative percentage variance	3	1	15
Residual percentage variance	Ambiguous	1	Ambiguous
Average eigenvalue	1	1	7
Parallel analysis	1	1	7
Embedded error function	2	No solution	No solution
Autocorrelation	2	No solution	2
PRESS	3	1	15
R ratio	2	2	5
Akaike information criterion	2	No solution	No solution
Minimum description length	2	No solution	No solution
VRE_{cov}	2	1	2
VRE_{cor}	2	1	2

5. Fault Detection Indices

Once a PCA model is developed by analyzing the historical data under normal operating conditions, new observation samples can be projected to the PC subspace to obtain their scores and loadings and compare them further with the control limits. Typically, the squared prediction error (SPE, also referred to as the Q statistic) and Hotelling's T^2 statistic are used to calculate control limits for multivariate cases. See [7] and [26] for discussions that interpret the role of SPE and T^2 in fault detection. If a new data set is collected, SPE measures the magnitude of the sample projection on the residual space (equation 8-2v), while Hotelling's T^2 statistic considers the variations in the principal component space (equation 8-2w) [7].

SPE is preferred over Hotelling's T^2 in fault detection applications because of their nature as described in the simple example presented earlier. In other words, use residual space to detect faults that are not explained by the model.

$$SPE = \| (\mathbf{I} - \mathbf{P}\mathbf{P}^{\mathbf{T}})x \|^2 \qquad (8\text{-}2v)$$

$$T^2 = x^T \mathbf{P}\mathbf{\Lambda}^{-1}\mathbf{P}^{\mathbf{T}}x \qquad (8\text{-}2w)$$

where:

$\mathbf{\Lambda}$ is a diagonal matrix: $\mathbf{\Lambda} = \mathrm{diag}(\lambda_1, \lambda_2, \ldots, \lambda_k)$.

The upper control limit of SPE is given by Jackson and Mudholkar [27] as follows:

$$SPE_\alpha = \theta_1 [\frac{c_\alpha \sqrt{2\theta_2 h_0^2}}{\theta_1} + 1 + \frac{\theta_2 h_0 (h_0 - 1)}{\theta_1^2}]^{\frac{1}{h0}} \qquad (8\text{-}2x)$$

where:

$$\theta_i = \sum_{j=k+1}^{n} \lambda_j^i, i = 1,2,3$$

and

$$h_0 = 1 - \frac{2\theta_1\theta_3}{3\theta_2^2} \qquad (8\text{-}2y)$$

k is the number of PCs retained, and α is the significance level, while c_α is the standard normal deviation corresponding to the upper 1-α percentile.

The upper control limits of Hotelling's T^2 statistic can be calculated by using the F-distribution [20]:

$$T_\alpha^2 = \frac{k(m-1)}{m-k} F_{k,m-k;\alpha} \qquad (8\text{-}2z)$$

where F is the F-distribution and m is the number of observations. As discussed earlier, the two indices measure different variations of the sample. SPE is preferred over Hotelling's T^2 in fault detection calculations. Besides the two general indices, several other metrics are proposed in the literature, including combined indices [28], global Mahalanobis distance [7], and Hawkins's statistic [20].

Different Types of PCA and Applications

Section 8-2 introduced the core idea and basic algorithms of PCA. Because PCA has been applied broadly, many variations have been developed to meet specific requirements. Table 8-2d lists some of the variations of PCA and their usage in chemical and biological applications.

Table 8-2d. PCA Variations and Their Applications in Chemical and Biological Process Monitoring

Processes	Method	Articles	Method Features
Time Invariant Continuous Process	PCA	Kresta et al. [29]	Reduce system dimension
Time Variant Continuous Process	Recursive PCA	Li et al. [30]	Adaptive model monitoring
Dynamic Continuous Process	Dynamic PCA	Ku et al. [31], Russell et al. [32]	Time correlation between variables is considered
Nonlinear Continuous Process	Nonlinear PCA	Dong and McAvoy [33]	The nonlinearity in the data is considered
Large-scale processes	Multiblock PCA	MacGregor et al. [34]	Decompose large matrix into smaller blocks
Multiscale Continuous Process	Multiscale PCA	Bakshi [35]	Deal with signals using different scales
Continuous and Batch Processes	Model-based PCA	Rotem et al. [11]	First-principle model is used to simulate the process
Batch process	Multiway PCA	Nomikos and McGregor [22, 36, 37]	Three-dimensional batch data is unfolded to two dimensions

8-3. Multiway PCA

In section 8-2, we gave some important definitions and algorithms for applying PCA to continuous processes. However, for batch and semibatch processes, there is no steady state and the historical trajectories usually contain considerable nonlinearity, so no effective online monitoring technique existed before 1995. Nomikos and MacGregor [22] [36] [37] originally introduced the basic ideas of Multiway PCA and PLS methods to monitor batch processes in real time. In this section, we provide the general ideas and computational steps for performing PCA on batch processes and introduce one variation of PCA (Multiway PCA).

Reference Trajectory

Batch processes usually exhibit some batch-to-batch variation because of the process variables deviate from their specified trajectories, errors occur during the charging of the materials, and disturbances arise from variations in impurities [22]. Besides the variations between batches, the trajectories have general similarities because of the kinetics and transport limits in the system. Fault detection extracts the system's intrinsic dynamics as a reference trajectory and then measures a sample trajectory's deviation with respect to the reference. If the deviation is out of the control limits, an alarm is triggered.

In some cases, the reference trajectory is used directly in fault detection and identification without performing the PCA step [38-40].

PCA of a Batch Process

The Multiway PCA procedure for monitoring and analyzing batch processes can be summarized in terms of the following two steps:

Step A: Building the PCA Model

1. Synchronizing batch trajectory (optional, see section 8-3)

2. Unfolding the normal operating condition's historical data $\overline{X}(I \times J \times K)$ into a two-dimensional array (figures 8-3a to 8-3c). I is the number of batches; J is the variable number, and K is the sampling time.

3. Normalizing the data.

4. Calculating the principal components and extracting score and loading matrices.

5. Obtaining the upper control limits of *SPE* and Hotelling's T^2.

Step B: Detecting Faults

The data preprocessing (first three) tasks in step B are the same as those in step A.

Offline test: calculate the score of a new batch at the end of a run, and compare it with the upper control limits. This procedure is usually used to check product quality.

Online test: calculate the score of a new batch at regular time intervals during the batch, and compare the results with upper control limits. An online test is used to perform real-time monitoring. If an alarm is triggered, the contribution plot, which shows the contribution of each variable to the scores of *SPE* and Hotelling's T^2 at a specific time, is used to diagnose faults. The algorithm for the online test depends on the specific unfolding method used. A detailed PCA algorithm for batch-wise unfolding is given in *PCA algorithm* later.

To perform the procedure just described, pay special attention to the following specific steps:

a) Historical data unfolding (step A.2) and normalization (step A.3)
The main difference between continuous process and batch process historical data is that the former is stored in a two-dimensional matrix, while batch data is usually stored in a three-dimensional matrix. Therefore, to perform PCA, the dataset should be unfolded to two dimensions. See Figures 8-3a, 8-3b, and 8-3c.

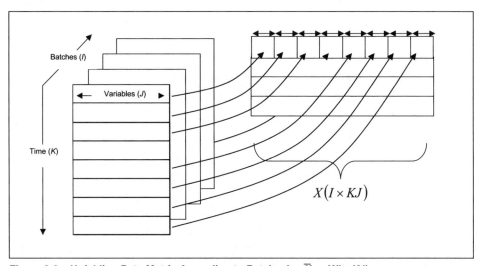

Figure 8-3a. Unfolding Data Matrix According to Batch-wise $\overline{X} \rightarrow X(I \times KJ)$

In step A.2, all the trajectories are assumed to have equal length. Usually, the historical data are stored in $\overline{X}(I \times J \times K)$, where I is the number of batches, J is the variable number, and K is the sampling time (observations). Recently, Kourti [41] has pointed out that the measurements may not be equally spaced and that some variables do not exist throughout the whole processes. These situations are variations of the general case (K is the same for each batch), but in this chapter J and K are assumed constant.

Van Sprang et al. [42] provided a critical evaluation for online batch processing and discussed different data-unfolding methods. There are

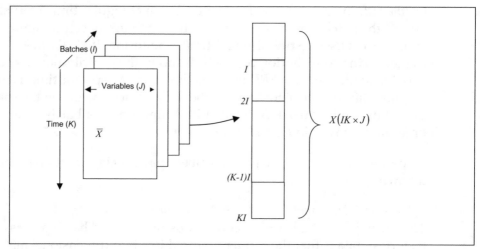

Figure 8-3b. Unfolding Data Matrix According to Variable-wise $\overline{X} \to X(IK \times J)$

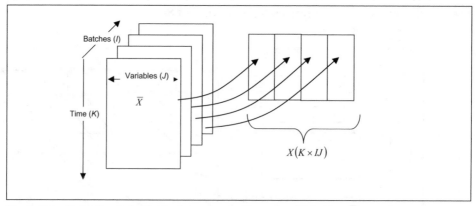

Figure 8-3c. Unfolding Data Matrix According to Time-wise $X(K \times IJ) \to \overline{X}$

three ways to unfold the data: batch-wise, variable-wise, and time-wise (figures 8-3a to figure 8-3c). Besides van Sprang et al., Lee et al. [43], Westerhuis et al. [44], and Kourti [41] have also compared the three different unfolding methods.

> *For batch process monitoring, it is recommended that you include batch time as a variable when performing PCA.*

The way by which unfolding method uses leads directly to the variation information that PCA extract. Batch-wise unfolding focuses on analyzing the differences among batches, variable-wise unfolding attempts to discover the variability between variables, and time-wise unfolding is used to extract the correlation among samples (at different times). The first two methods are widely used in batch monitoring, and we will discuss them in detail here. After unfolding, the mean-centering and rescaling steps (step A.3) are carried out as described in section 8-2 (data preprocessing).

When batch-wise unfolding is used, the mean trajectory reflects the dynamics of the system. After processing, the residual is the deviation of the specific trajectory from the average dynamic profile of the process. The dynamic behavior of the batch process is removed by this method, which is a big advantage, but the row vector data will not be complete until the end of a batch (figure 8-3a). In PCA online monitoring, the score and loading calculations need a complete data set. Thus, one has to predict the future values for the whole batch [22] [36], which is time consuming and can add uncertainty, especially during a batch's initial period. The variable-wise method discovered by Wold et al. [45] does not have this problem because only the current time data matrix ($I{\times}K$) is needed for each time (a small data matrix shown in figure 8-3b). The shortcoming of the variable-wise approach is that the system dynamics are still included in the data set after preprocessing. Thus, the calculated loadings will contain the correlations between variables, with time dependency included [41] [43]. Furthermore, Westerhuis et al. [44] found that variable-wise analysis offers little benefit to monitoring, since it focuses on the wrong source of variations in the data.

Recently, Lee et al. [43] combined the two batch-wise and variable-wise methods, which we called "hybrid-wise unfolding," and applied it successfully to a bioreactor fault detection case. At first, the data set is unfolded batch-wise, and the mean-centering and scaling steps are performed. After that, the data is rearranged to variable-wise. In this way, the time dependency is canceled, and future data prediction is also avoided.

Besides data unfolding methods, several other methods, such as parallel factor analysis (PARAFAC) [46] [47] and Tucker models [18] [48], analyze three-dimensional data directly. Westerhuis et al. [44] also found that unfolding is preferred to other direct three-dimensional analysis methods. However, Chiang et al. [18] suggested unfolding method gives similar results to three-dimensional analyze method. After all, after unfolding, PCA is able to capture most of the variation of the system and fits the data best.

b) PCA Algorithm
After data preprocessing, the data set is similar to what we have in steady-state continuous processes, so the following procedures in step A are the same as in continuous process. The upper control limits are calculated by equations 8-2x to 8-2z. In step B, the offline monitoring is composed of data preprocessing (as in step A) and data projection (as in a continuous

process [section 8-2]). The online monitoring approach is different. Using batch-wise unfolding as an example, the main steps are as follows:

1.   Perform the normalization steps as in step A, and analyze the new incomplete batch data ($k < K$).

2.   Calculate the new score of the batch using the "PCA projection" method (the third method for anticipating the future data is given in [22]), which also takes the missing data into account:

$$t_k = (\mathbf{P}_k^T \mathbf{P}_k)^{-1} \mathbf{P}_k^T x_{new,k} \tag{8-3a}$$

3.   The projection of x on principal and residual spaces is:

$$\hat{x}_{new} = t_{new} \mathbf{P}^T \tag{8-3b}$$

$$\tilde{x} = x_{new} - \hat{x}_{new} \tag{8-3c}$$

Equations 8-3b and 8-3c are similar to equations 8-2t and 8-2u. $\mathbf{P}$ is the calculation result of loading matrix from step A.

4.   Calculate the scores of Hotelling's T^2 and SPE by:

$$SPE_{new}(k) = \sum_{j=1}^{J} \tilde{x}_{new,jk}(k)^2 \tag{8-3d}$$

$$T_{new}^2 = t_{new}(S)^{-1} t_{new}^T \tag{8-3e}$$

where $\mathbf{S}$ is the covariance matrix of the score calculated in step A. These two equations are variations of equations 8-2v and 8-2w. Step 4 is repeated for every new sample.

For batch-wise unfolding, future data anticipation is needed, while for variable-wise unfolding, this step is skipped, as discussed previously. Lee et al. [43] give a detailed algorithm scheme of variable-wise PCA algorithm.

5.   Compare the results of SPE and T^2 with the upper control limits.

The steps after data preprocessing are nearly universal for all types of PCA in batch processes. Thus, in later sections, the preprocessing steps are discussed in detail.

Batch Data Synchronization
Even for successful batches, certain batch operating steps may require different durations from batch to batch. If principal component analysis

(PCA) is used directly without synchronizing trajectories, the result can be misleading. For a bioreactor, the oxygen concentrations, fermenter temperature, and other variables may vary around their normal profiles. Unsteady behavior will affect the duration of each stage and result in different batch lengths. Furthermore, synchronization between different batches is a necessary assumption in order to perform batch data monitoring [22] [37].

Techniques such as dynamic time warping (DTW) [49], correlation optimized warping [50], indicator variable [36] [51], and data interpolation [9] have been introduced to synchronize batch profiles. In some cases, the indicator variable approach does not work [41]. Simple interpolation within the data can sometimes be employed, but it does not guarantee synchronization (the batches may still reach completion at different times). Dynamic time warping is more complicated, but it captures the dynamics, matches patterns for batches of different lengths, and has broad application. DTW, first used in speech recognition, was introduced for batch data analysis by Kassidas et al. in 1998 [49]. After that, the DTW method was successfully applied to process monitoring in industrial emulsion polymerization [49], spectroscopic profile analysis [50], semiconductor production monitoring [9], chromatographic data alignment [52], and other applications. Therefore, DTW is used in this chapter.

The core idea behind DTW is to align batches of different durations by finding the minimum accumulated path length between two trajectories.

Ramaker et al. [53] and Pravdova et al. [50] give clear explanations of how DTW works. We will use a simple example to demonstrate the algorithm, and later the method will be used on a bioreactor fermenter case. Suppose there are two trajectories X_1, X_2 of different length and each trajectory has two variables:

$$X_1 = \begin{bmatrix} 0 & 0.5 & 2 & 1.9 & 4 \\ 2 & 3 & 4 & 3.9 & 5 \end{bmatrix}$$

$$X_2 = \begin{bmatrix} 0 & 0.1 & 0.5 & 1.9 & 4 & 4.1 \\ 1.9 & 2 & 3.1 & 4 & 5.5 & 5 \end{bmatrix}$$

X_1 is a 2×5 matrix while X_2 is 2×6; each row represents a process variable. Figure 8-3d shows the optimal warping path between the two trajectories. The line represents the optimal path, and the coordinates of small circles represent the synchronized point between two trajectories. For example,

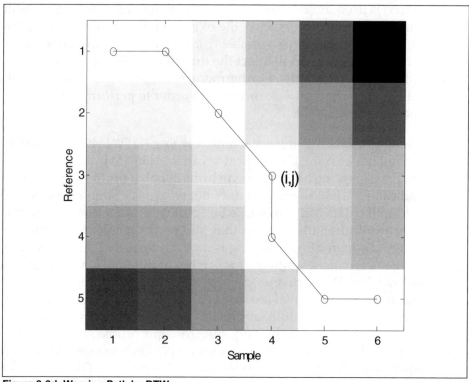

Figure 8-3d. Warping Path by DTW

i=4 and j=3 means that the fourth point in sample trajectory (X_2 [1.9; 4]) should be synchronized with the third point in reference trajectory (X_1 [1.9; 3.9]), which appears to make sense in figure 8-3d.

For implementing PCA off line, DTW can be carried out as described earlier since all data is available. However, for online cases, no complete trajectory is available, so we need to decide how to align current batch time with the reference trajectory. Kassidas et al. [49] suggest a way to do this aligning that is similar to the offline case. This step is carried out whenever a new observation is taken; thus, the computational load could be higher. However, with the development of personal computer, the alignment is calculated very quickly compared to the interval between observations. For a detailed online algorithm of DTW, see [49].

8-4.   Model-based PCA (MB-PCA)

MB-PCA was first developed by Wachs and Lewin [54] in 1998 with the aim of removing the process's dynamic behavior and nonlinearity without performing data synchronization. This method has been successfully applied to monitor many processes, including a batch reactor with cooling jacket [54], the Tennessee Eastman simulation of a plant-wide process [54], a three-stage ethylene compressor [11], and an NMOS fabrication process [10].

Algorithm of MB-PCA
Though there are many similarities between Multiway PCA and MB-PCA, in this section we focus on their differences. Because biological processes are generally run in a batch mode and a batch process is more complex than a continuous process, we discuss a batch process example in this section.

In Multiway PCA, new sample data is compared with an existing data set that consists of normal operating condition trajectories. If the deviation is too large, an alarm is triggered. As we pointed out in section 8-3, to perform Multiway PCA, all the trajectories must be synchronized first. However, methods such as DTW will reduce the system's ability to detect faults as noted earlier.

In MB-PCA, a first-principle model is used to describe the nonlinearity and dynamics of normal operating conditions. The sample data is compared with the calculated data using a first-principle model, and PCA is performed on the residual between the model predictions and the data (figure 8-4a). If the first-principle model is perfect, no dynamics will be left in the residual data set.

As a result, the residual vectors of every batch can be adjusted to have the same length without having to perform complex calculations. Then, the residual data matrix is unfolded ($E(I \times KJ)$ etc.) and scaled according to:

$$\tilde{e}_{ij} = (e_{ij} - e_{j,mean}) / \sigma(e_j)$$
(8-4a)

Finally, PCA is performed on the unfolded data, as described in section 8-3.

However, if the first-principle model is imperfect, the residuals will still include dynamic effects and can lead to incorrect monitoring results.

In summary, MB-PCA can be used when a first-principle model is accurate enough to describe the dynamics of a process. It can effectively enhance the performance of PCA [10] [11] [54]. However, for large-scale systems in which the dynamics may not be well known, the MB-PCA can lead to wrong decisions.

Super Model-based PCA
As mentioned in section 8-4, MB-PCA is very useful when the process can be accurately described by a first-principle model. However, in an industrial environment this assumption is generally not realistic.

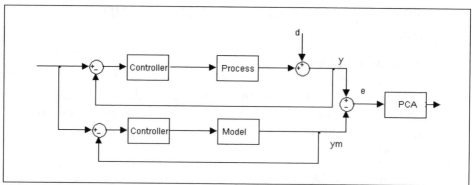

Figure 8-4a. MB-PCA Flow Scheme

McPherson et al. [55] added a residual modeling stage to modify MB-PCA so the structural dynamics in the residuals would be removed.

Figure 8-4b is a schematic of SMB-PCA. The residual data (E matrix in MB-PCA) and output from the plant are considered in the error model. The error model is used to predict the remaining structure in E, and the structural part is subtracted again from the residual matrix (E). Then PCA or other monitoring techniques are utilized. So, when using a batch process as an example, the main steps in SMB-PCA are as follows:

Step A: Building the PCA Model

1. The values predicted by the model are subtracted from the raw data, and data preprocessing is performed, resulting in a residual matrix (E).

2. The structural part in the residual matrix is estimated using the error model.

3. Estimated error values are subtracted from the residual matrix to obtain an unstructured residual matrix ($\hat{E}$).

4. PCA is applied to extract information from $\hat{E}$.

5. Upper control limits of *SPE* and Hotelling's T^2 are obtained.

Step B: Monitoring
The data preprocessing tasks are the same as those in step A and in data projection. Score calculations are the same as those in Multiway-PCA.

Residual model-building is a key step in SMB-PCA. McPherson et al. [26] tested five ways to model the structured residuals: partial least squares (PLS), auto-regressive with exogenous input (ARX), dynamic PLS, dynamic nonlinear PLS, and dynamic canonical correlation analysis (CCA). In McPherson et al.'s work, an exothermic batch reactor process (the same example that Wachs and Lewin used to develop MB-PCA) was

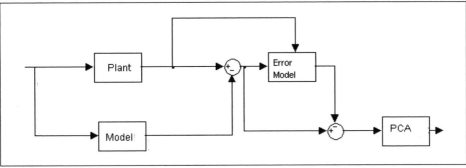

Figure 8-4b. SMB-PCA Flow Scheme

monitored. The results suggested that all the dynamic error models are excellent at detecting faults, whereas the PLS with SMB-PCA does not perform well because the variables in the residual matrix mainly have dynamic nonlinear correlation.

It can be seen from table 8-2d and the foregoing discussion that PCA has many variations, and most of them have specific applications. Multiway PCA and MB-PCA are attractive for batch processes because they can be applied easily and offer good monitoring capability. Figure 8-4c shows a flow chart for choosing a monitoring technique for a batch system. Table 8-4a compares the shortcomings and advantages of each method and gives guidance on the choice of methods. For a simple system in which the physical model is exactly known, MB-PCA is recommended because the synchronization and unfolding steps can be skipped. For a large system for which physical information is sufficient but limited, SMB-PCA may perform better, offering a good balance between computational loading and fault detection ability.

Multiway-PCA is a more universal method than the other two because it does not require a physical model but uses normal operation data; hence, it can be implemented with relative ease and speed.

Table 8-4a. Multiway-PCA, MBPCA, and SMBPCA Comparison

	Multiway-PCA	**MB-PCA**	**SMB-PCA**
Physical model	Not needed	Accurate physical model information	Basic physical model information
Synchronization (such as DTW)	Suggested	Not needed	Not needed
Different unfolding method	Further treatment (multiple options)	Does not matter	Does not matter
Applicable system	Does not matter	Simple system	Does not matter

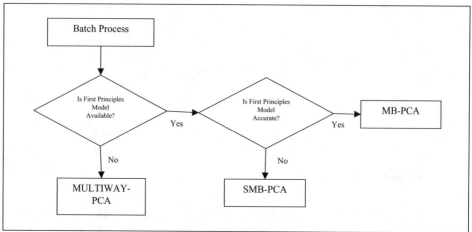

Figure 8-4c. Flow Chart of PCA Variation for Batch Process

8-5.   Fault Detection

Bioreactor simulation
In this section we use a detailed nonlinear dynamic model to simulate
control of a batch penicillin fermentation process. The high-fidelity
fermentation model used here is proposed by Birol et al. [56]. The batch
fermentation process consists of seventeen control modules (the CD
accompanying this book provides for more detail on the model). The
simulated bioreactor volume is 1000L, and the time scale is one thousand
times faster than real time. Figure 8-5a shows one process module of the
system. Small variations are added to the simulation input variables in
order to mimic real industrial plants. In this way, all the normal batches
will fluctuate around a mean trajectory.

The simulated data can be directly read from the OPC server, which is
very convenient for online monitoring and control. There is also a
historical data server that can be used to save historical data for future use.
Figure 8-5b shows the trends of some important process variables that
change during the run. PID controllers are used to control dissolved
oxygen, substrate concentration and broth pH.

PCA Application
According to McMillan, nine important process variables are used in PCA.
These include process inputs as base reagent flow (kg/sec), vent flow (kg/
sec); the process output variables include head pressure (kPa), biomass
concentration (g/L), product concentration (g/L), biomass growth rate
(kg/hr), and current product yield (% max) and batch time (hour),

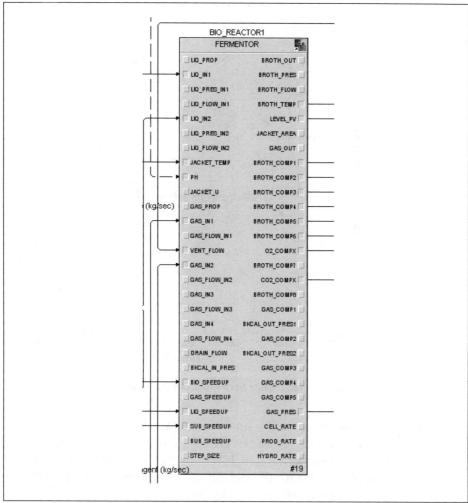

Figure 8-5a. Process Module of Bioreactor

> *Which monitored process variables are selected for PCA is an important decision; the choice will depend on the user's experience in this field. You need to monitor the right variables in order to identify abnormal behavior.*

Twenty normal batch profiles generated by the simulator described earlier are used to build a PCA model and to calculate upper control limits (α=0.99) of SPE and T^2 control charts. The Multiway PCA calculation results are shown in this section, and both online and offline tests are applied. For the offline test, the batch-wise unfolding method is used. For the online case, both variable-wise and hybrid-wise unfolding methods are applied. Related algorithms are listed in section 8-3 [43]. Dynamic time warping (DTW) is used to synchronize different batch lengths. For a similar process, four principal components were retained in the PCA

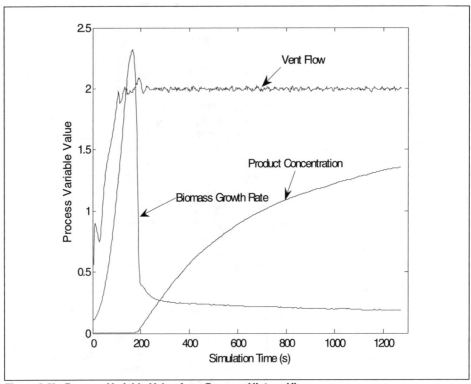

Figure 8-5b. Process Variable Value from Process History View

model [43]. In our case, for an offline test, six PCs are retained, capturing 54 percent of the variation. Two PCs are used for online testing with variable-wise unfolding, which retains 84.5 percent of the total variation. For the online test with the hybrid-wise unfolding case, two PCs capture 78.9 percent of the information.

> *The DTW method is more helpful for real industrial situations. For our simulation, if the batches are relatively well synchronized, the DTW step can be bypassed.*

Two types of faults are tested by Multiway PCA: (1) pH sensor failure at time zero and (2) initial biomass seed overload (from 1kg to 1.5 kg). To compare and test the PCA approach, a normal batch was also tested.

With respect to diagnosing abnormal batches, table 8-5a lists the offline test results, and figure 8-5c shows the offline *SPE* control chart of normal and abnormal cases. It can be seen that PCA treats the normal batch data as a good batch, and both faults are detected successfully. As mentioned in previous sections, *SPE* is focusing on the residual space, which measures the deviation between good batches (those used to build PCA model) and tested ones. The results show that with six PCs, the residual space can correctly distinguish a faulty batch from good ones.

Table 8-5a. PCA Offline Test Results

	T^2	SPE	T^2 99% UCL	SPE 99% UCL
Normal batch	0.09	0.67	23.8	0.74
pH sensor failure	1.63	3.08 *	23.8	0.74
Initial biomass seed overload	0.40	1.34 *	23.8	0.74

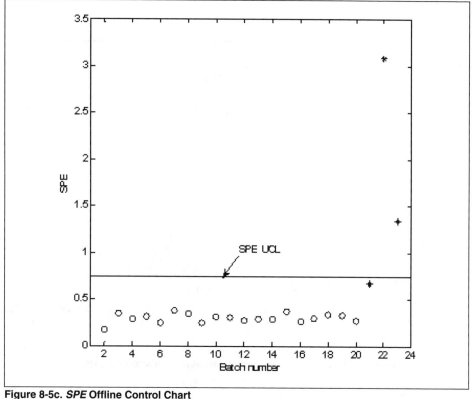

Figure 8-5c. *SPE* **Offline Control Chart**
Note: 'o' represents batches used in the PCA model, and '*' represents batches tested for abnormality.

As mentioned in section 8-3, offline PCA is typically used to check the quality of the final product and is performed at the end of every batch. To achieve the purpose of monitoring, online PCA should be applied. In this section, variable-wise and hybrid-wise unfolding methods [43] are used. An important feature of these methods is that they do not need to estimate unfinished batch profile because only current data is used (section 8-3). Figures 8-5d to 8-5i show the control charts for the online monitoring of T^2. Both approaches classify normal batches correctly. For the pH sensor failure test, for both methods T^2 scores exceed the UCL at an early stage, which means that the fault is detected rapidly. However, for the case of initial biomass overload, the two methods give different answers (figures 8-5f and 8-5i). The variable-wise unfolding results indicate that the batch is good, but the hybrid-wise unfolding shows that the batch has abnormality at the beginning. In section 8-3, we pointed out that variable-wise unfolding cannot completely remove system dynamics from a raw data set, and this fact will influence the performance of PCA, as the example demonstrates.

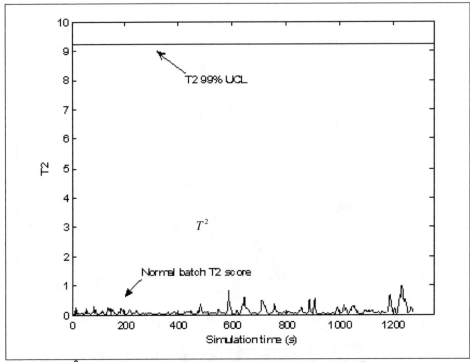

Figure 8-5d. T^2 Score of Another Normal Batch with Variable-Wise Unfolding

Hybrid-wise unfolding works better than variable-wise because when performing mean centering the process dynamic is subtracted out.

In offline monitoring, only one value is calculated for every batch (figure 8-5c), which indicates the quality of the product. In real applications, a UCL is calculated to determine when the product should be discarded. For online monitoring, current process variable values are measured at regular time intervals and analyzed by PCA or other MSPC methods. If the computed results exceed the corresponding UCL, then further fault diagnosis and analysis should be performed to isolate and correct the process problem. It has been shown that online PCA can detect abnormalities early in the batch process, normally giving you enough time to take remedial action.

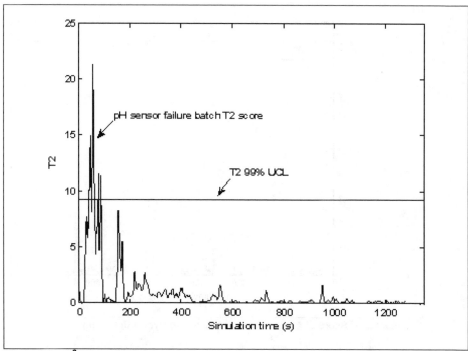

Figure 8-5e. T^2 **Score of pH Sensor Failure Batch with Variable-Wise Unfolding**

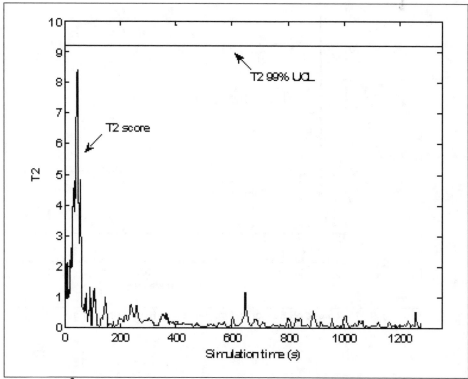

Figure 8-5f. T^2 **Score of Initial Biomass Overload Batch with Variable-Wise Unfolding**

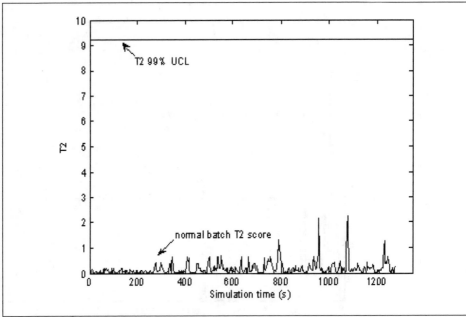

Figure 8-5g. T^2 **Score of Another Normal Batch with Hybrid Unfolding**

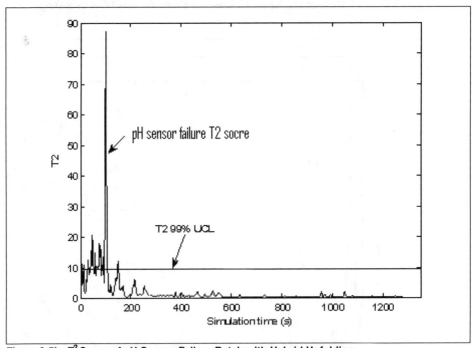

Figure 8-5h. T^2 **Score of pH Sensor Failure Batch with Hybrid Unfolding**

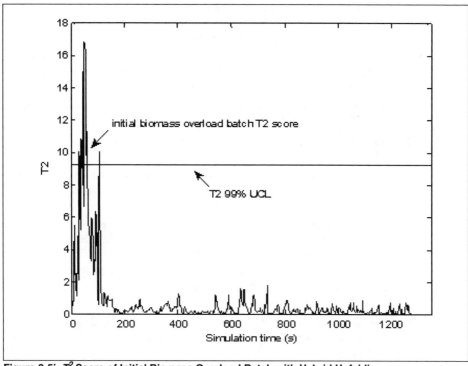

Figure 8-5i. T^2 **Score of Initial Biomass Overload Batch with Hybrid Unfolding**

References

8-1.   Kourti, Theodora, "The Process Analytical Technology Initiative and Multivariate Process Analysis, Monitoring and Control." *Analytical and Bioanalytical Chemistry* 384 (2006): 1043-48.

8-2.   Chiang, Leo H., Russell, Evan L., and Braatz, Richard D., *Fault Detection and Diagnosis in Industrial Systems*. Springer Verlag, 2001.

8-3.   MacGregor, John F., and Kourti, Theodora, "Statistical Process Control of Multivariate Processes." *Control Engineering Practice*, 3 (1995): 403-14.

8-4.   Lee, Jong-Min, "Statistical Process Monitoring Based on Independent Component Analysis and Multivariate Statistical Methods." Ph.D. dissertation, Department of Chemical Engineering, Pohang University of Science and Technology, Pohang, South Korea, 2004.

8-5.   Mason, Robert L., and Young, John C., *Multivariate Statistical Process Control with Industrial Applications*. Philadelphia: ASA-SIAM, 2002.

8-6.   Valle-Cervantes, Sergio, "Plant-wide Monitoring of Processes under Closed-loop Control." PhD Dissertation, Chemical Engineering Department, University of Texas at Austin, Austin, 2001.

8-7.   Qin, S. Joe, "Statistical Process Monitoring: Basics and Beyond." *Journal of Chemometrics*, 17 (2003): 480-502.

8-8.   Hokuldsson., Agnar, "PLS Regression Methods." *Journal of Chemometrics*, 2 (1988): 211-28.

8-9.   Cherry, Gregory A., and Qin, S. Joe, "Batch Synchronization and Monitoring Using Data Interpolation and Dynamic Time Warping with Principal Component Analysis." Presentation at TWMCC meeting, Austin, Texas, 2005.

8-10.   Lachman-Shalem, Sivan, Haimovitch, Nir, Shauly, Eitan N., and Lewin, Daniel R., "MBPCA Application for Fault Detection in NMOS Fabrication." *IEEE Transactions on Semiconductor Manufacturing,* 15 (2002): 60-70.

8-11.   Rotem, Y., Wachs, A., and Lewin., D. R., "Ethylene Compressor Monitoring Using Model-Based PCA." *AIChE Journal,* 46 (2000): 1825-36.

8-12.   Kourti., Theodora, "Monitor." *Chemometrics and Intelligent Laboratory Systems,* 76 (2005): 215-20.

8-13.   Miletic, Ivan, Quinn, Shannon, Dudzic, Michael, Vaculik, Vit and Champagne. "An Industrial Perspective on Implementing Online Applications of Multivariate Statistics." *Journal of Process Control,* 14 (2004): 821-36.

8-14.   Kourti, Theodora, Lee, Jennifer, and MacGregor, John F., "Experiences with Industrial Applications of Projection Methods for Multivariates Statistical Process Control." *Computers and Chemical Engineering,* 20 (1996): S745-S750.

8-15.   Albert, Sarolta and Kinley, Robert D., "Multivariate Statistical Monitoring of Batch Processes: An Industrial Case Study of Fermentation Supervision." *Trends in Biotechnology,* 19 (2001): 53-62.

8-16.   Lennox, Montague, G. A., Hiden, H. G., Kornfeld, G., and Goulding, P. R., "Process Monitoring of an Industrial Fed-batch Fermentation." *Biotechnology and Bioengineering,* 74 (2001): 125-35.

8-17.   Lopes, J. A., Menezes, J. C., Westerhuis, J. A., and Smilde, A. K., "Multiblock PLS Analysis of an Industrial Pharmaceutical Process." *Biotechnology and Bioengineering,* 80, (2002): 419-27.

8-18.   Chiang, L. H., Leardi, R., Pell, R. J., and B., Seasholtz M., "Industrial Experiences with Multivariate Statistical Analysis of Batch Process Data." *Chemometrics and Intelligent Laboratory Systems* 81 (2005): 109-19.

8-19.   Jolliffe, Ian T., *Principal Component Analysis.* 2d ed. Springer-Verlag, 2002.

8-20.   Jackson, J. Edward, *A User's Guide to Principal Components,* John Wiley, 2003.

8-21.   Mathworks, Inc., MATLAB, 2006.

8-22.   Nomikos, Paul, and MacGregor, John F., "Multivariate SPC Charts for Monitoring Batch Processes." *Technometrics,* 37 (1995): 41-59.

8-23.   Strang, Gilbert, *Linear Algebra and Its Applications,* 2d ed. New York: Academic Press, 1980.

8-24.   Geladi, Paul, and Kowalski, Bruce R., "Partial Least Squares Regression: A Tutorial." *Analytica Chimica Acta,* 185 (1986): 1-17.

8-25.   Valle, Sergio, Li, Weihua, and Qin, S. Joe, "Selection of the Number of Principal Components: The Variance of the Reconstruction Error Criterion with a Comparison to Other Methods." *Industrial & Engineering Chemistry Research,* 38 (1999): 4389-401.

8-26.   Wise, Barry M., and Gallagher, Neal B., "The Process Chemometrics Approach to Process Monitoring and Fault Detection." *Journal of Process Control,* 6 (1996): pp. 329-348, 1996.

8-27.   Jackson, J. Edward, and Mudholkar, Govind S., "Control Procedures for Residuals Associated with Principal Component Analysis." *Technometrics,* 21 (1979): 341-49.

8-28.   Yue, H. Henry, and Qin, S. Joe, "Reconstruction-based Fault Identification Using a Combined Index." *Industrial & Engineering Chemistry Research*, 40 (2001): 4403-4414, 2001.

8-29.   Kresta, J.V., MacGregor, J. F., and Marlin, T.E., "Multivariate Statistical Monitoring of Process Operating Performance." *Canadian Journal of Chemical Engineering*, 69 (1991): pp. 35-47.

8-30.   Li, Weihua, Yue, H. Henry, Valle-Cervantes, Sergio, and Qin, S. Joe, "Recursive PCA for Adaptive Process Monitoring." *Journal of Process Control*, 10 (2000): 471-86.

8-31.   Ku, Wenfu, Storer, Robert H., and Georgakis, Christos, "Disturbance Detection and Isolation by Dynamic Principal Component Analysis." *Chemometrics and Intelligent Laboratory Systems*, 30 (1995): 179-96.

8-32.   Russell, Evan L., Chiang, Leo H., and Braatz, Richard D., "Fault Detection in Industrial Processes Using Canonical Variate Analysis and Dynamic Principal Component Analysis." *Chemometrics and Intelligent Laboratory Systems*, 51 (2000): 81-93.

8-33.   Dong, Dong, and McAvoy, Thomas J., "Nonlinear Principal Component Analysis Based on Principal Curves and Neural Networks." *Computers and Chemical Engineering* 20 (1996): 65-78.

8-34.   MacGregor, John F., Jaeckle, Christiane, Kiparissides, Costas, and Koutoudi, M., "Process Monitoring and Diagnosis by Multiblock PLS Methods." *AIChE Journal*, 40 (1994): 826-38.

8-35.   Bakshi, Bhavik R., "Multiscale PCA with Application to Multivariate Statistical Process Monitoring." *AIChE Journal*, 44 (1998): 1596-1610.

8-36.   Nomikos, Paul, and MacGregor, John F., "Monitoring of Batch Processes Using Multi-way Principal Component Analysis." *AIChE Journal*, 40 (1994): 1361-75.

8-37.   Nomikos, Paul, and MacGregor, John F., "Multi-way Partial Least Squares in Monitoring Batch Process." *Chemometrics and Intelligent Laboratory Systems*, 30 (1995): 97-108.

8-38.   Zhang, Yang, and Edgar, Thomas F., "Application of MB-PCA and Golden Batch to Bioreactor Monitoring and Control." Presentation at TWMCC meeting, Austin, 2006.

8-39.   "Automation Rockwell Improves Consistency, Quality and Productivity." http://www.rockwellautomation.com/solutions/process/improve.html

8-40.   Emerson, David, "Production Performance Ratings Measure Effectiveness of Batches." At *ISA's online Technical Information and Communities; http://www.isa.org/Template.cfm?Section=Communities&template=/TaggedPage/DetailDisplay.cfm&ContentID=40681*, 2004.

8-41.   Kourti, Theodora, "Multivariate Dynamic Data Modeling for Analysis and Statistical Process Control of Batch Processes, Start-ups and Grade Transitions." *Journal of Chemometrics*, 17 (2003): 93-109.

8-42.   Sprang, Eric N. M. van, Ramaker, Henk-Jan, Westerhuis, Johan A., Gurden, Stephen P., and Smilde, Age K., "Critical Evaluation of Approaches for Online Batch Process Monitoring." *Chemical Engineering Science*, 57 (2002): 3979-91.

8-43.   Lee, Jong-min, Yoo, Chang Kyoo, and Lee, In-Beum, "Enhanced Process Monitoring of Fed-batch Penicillin Cultivation Using Time-varying and Multivariate Statistical Analysis." *Journal of Biotechnology*, 110 (2004): 119-36.

8-44.   Westerhuis, Johan A., Kourti, Theodora, and MacGregor, John F., "Comparing Alternative Approaches for Multivariate Statistical Analysis of Batch Process Data." *Journal of Chemometrics*, 13 (1999): 397-413.

8-45.   Wold, Svante, Kettaneh, Nouna, Friden, Hakan, and Holmberg, Andrea, "Modelling and Diagnostics of Batch Processes and Analogous Kinetic Experiments." *Chemometrics and Intelligent Laboratory Systems*, 44 (1998): 331-40.

8-46.   Smilde, Age K., and Doornbos, Durk A., "Three-way Methods for the Calibration of Chromatographic Systems: Comparing PARAFAC and Three-way PLS." *Journal of Chemometrics*, 5 (1991): 345-60.

8-47.   Bro, Rasmus, "PARAFAC, Tutorial and Applications." *Chemometrics and Intelligent Laboratory Systems*, 38 (1997): 149-71.

8-48.   Smilde, Age K., "Three-way Analysis: Problems and Prospects." *Chemometrics and Intelligent Laboratory Systems*, 15 (1992): 143-57.

8-49.   Kassidas, Athanassios, MacGregor, John F., and Taylor, Paul A., "Synchronization of Batch Trajectories Using Dynamic Time Warping." *AIChE Journal* 44 (1998): 864-75.

8-50.   Pravdova, V., Walczak, B., and Massart, D. L., "A Comparison of Two Algorithms for Warping of Analytical Signals." *Analytica Chimica Acta* 456 (2002): 77-92.

8-51.   Kourti, Theodora, Lee, Jennifer and Macgregor, John F., "Experiences with industrial Applications of Projection Methods for Multivariate Statistical Process Control." *Computers and Chemical Engineering* 20 (1996): 745-50.

8-52.   Tomasi, Giorgio, Berg, Frans van den, and Andersson, Claus, "Correlation Optimized Warping and Dynamic Time Warping as Preprocessing Methods for Chromatographic Data." *Journal of Chemometrics* 18 (2004): 231-41.

8-53.   Ramaker, Henk-Jan, Sprang, Eric N. M. van, Westerhuis, John A., and Smilde, Age K., "Dynamic Time Warping of Spectroscopic BATCH Data." *Analytica Chimica Acta* 498 (2003): 133-53.

8-54.   Wachs, A., and Lewin, D. R., "Process Monitoring Using Model-based PCA." *Proceedings of IFAC Symposium on Dynamics and Control of Process Systems*, Corfu, 1998: 86-92.

8-55.   McPherson, Lindesay, Morris, Julian, and Martin, Elaine, "Super Model-based Techniques for Batch Performance Monitoring." In *European Symposium on Computer Aided Process Engineering, 12*, 2002.

8-56.   Birol, Gulnur, Undey, Cenk, and Cinar, Ali, "A Modular Simulation Package for Fed-batch Fermentation: Penicillin Production." *Computers and Chemical Engineering, 26* (2002): 1553-65.

Appendix A:
Definition of Terms

Appendix A

Definition of Terms

Terms Used Interchangeably for Process Control

attenuation – filtering

dead band – backlash

control interval – cycle time – execution time

delay time – dead time

derivative action – rate action

derivative time – rate time

digital filter – process variable filter – signal filter

estimator – inferential measurement – intelligent sensor

integral action – reset action

integral time – reset time

lag time – time constant

open loop gain – plant gain – process gain

open loop time constant – plant time constant – process time constant

resolution – sensitivity – sticktion – stick-slip

steady state gain – static gain

total loop dead time – system dead time – plant dead time – process dead time

upset – disturbance – load

Definitions of Process Control Terms

See ANSI/ISA-51.1 Process Instrumentation Terminology for a comprehensive list of terms used in automation systems.

action – relative direction of change in output for a change in input ("direct" for changes in same direction and "reverse" for changes in opposite direction)

A/D – analog-to-digital converter typically used to convert an analog input to its digital value with a resolution of 0.05 percent for a 12-bit converter with one sign bit

attenuation – reduction in amplitude of an oscillation between two operating points or two frequencies as a result of smoothing by a filter time constant in the measurement or by process time constant from a mixed volume.

automatic mode – PID controller mode to enable its set point to be locally adjusted by the operator (also known as LOCAL and AUTO mode)

backlash – see dead band

cascade mode – secondary PID controller mode to enable its set point to be remotely adjusted by the output of a primary controller (also known as REMOTE and CAS mode)

closed loop time constant – time it takes for the process variable to reach 63 percent of its change after a time delay for a step change in the controller set point when the controller is in automatic. The controller tuning must provide a non-oscillatory response for this term to be valid. For integrating or runaway processes, the controller gain must be large enough for the control action to bend over the response within the scale range.

compression – smallest change in value of a variable that is stored by a computer or data historian (must not exceed the repeatability of the CV or the resolution of the MV)

control chart – a chart that measures the performance of a process and indicates when a process deviates from expected values

control interval – time interval between successive executions for digital devices (same as scan or update time and the inverse of scan rate or frequency)

controlled variable – a controller input to be kept at a set point. It is often a process output such as concentration, pH, pressure, temperature, or level but is normally in percent for the control algorithm)

controller action – relative direction of change in controller output for a change in the controlled variable ("direct" for same direction and "reverse" for opposite direction)

controller gain – tuning parameter multiplier for the proportional mode (dimensionless). It is the multiplier for the integral and derivative modes in most industrial PI and PID controllers. It is the 100 percent divided by the proportional band.

controller output – the output of a PID controller (normally in percent for the control algorithm)

controller tuning – adjustment of the proportional, integral (reset), and derivative mode settings for a PID controller

cycle time – see control interval

D/A – digital-to-analog converter typically used to convert a digital value to an analog output with a resolution of 0.05 percent for a 12-bit converter with one sign bit

data unfolding – a term used in batch data processing that indicates unfolding the three dimensional batch data into two dimensions

DDC mode – PID controller direct digital control mode to enable its output to be adjusted by a sequence, interlock, or a computer (also known as ROUT mode)

dead band – minimum change in the input for a reversal of direction that will cause a change in the output (important for a control valve and a manipulated variable)

dead time – time it takes for the process variable to get out of the noise band after a change in the manipulated variable (same as time delay)

delay time – see dead time

derivative action – controller output changes that are proportional to the rate of change of the error or the controlled variable (derivative mode of a PID controller)

derivative time – see rate time

digital filter – a first order time constant in a digital device to attenuate noise

Distributed Component Object Model (DCOM) – a protocol that enables software components to communicate directly over a network in a reliable and secure way

disturbance – a change in a process input or operating condition

dynamic gain – ratio of the output amplitude to a sinusoidal input amplitude after transients have decayed (also known as amplitude ratio and magnitude ratio)

estimator – online calculation (intelligent sensor or inferential measurement) of a process variable (e.g., composition or quality) corrected by a field or lab measurement

execution time – see control interval

feedback control – a control algorithm to reduce the error between the set point and the controlled variable (most often a PID or model predictive controller algorithm is used)

feedforward control – a computation of the manipulated variable from a measurement of the disturbance (most often corrected by a PID or model predictive controller)

filter time – time constant of a signal filter that is usually applied to the PV but can be inserted anywhere in the configuration as a function block (seconds or minutes)

filtering – see attenuation

final element – device used to adjust the manipulated variable (typically it is a control valve but it can be a variable speed drive, compressor vane, or an electrical heater)

hotelling's T^2 control chart – a multivariate process control chart in which the raw data are projected to the principal component subspace

inferential measurement – see estimator

integral action – controller output changes that are proportional to the integral of the error (integral mode of PID controller)

integral time – see reset time

integrating process – a process that ramps at a constant rate for a change in the manipulated variable when the controller is in manual if there are no disturbances

integrating process gain – percent change in ramp rate per percent change in controller output (%/sec/%).

intelligent sensor – see estimator

I/P – current to pneumatic converter mounted on the valve to convert a controller output to a pneumatic signal for the actuator (often integrated with positioner)

limit cycle – continuous oscillation of nearly equal amplitude that persists for a variety of controller gains (caused by a resolution limit or a nonlinearity such as stick-slip and pH)

loading matrix – P matrix in Eq. 8-2q; each column in the P matrix indicates the relationship between the columns of raw data (process variables for a continuous process)

manipulated variable – variable adjusted by a controller output (most often it is a process input such as flow but it can be the set point of a process variable for cascade control) (normally in process engineering units).

manual mode – operator adjusts the controller output directly (PID algorithm is suspended but will provide a bumpless transfer to automatic, cascade, or supervisory control)

matrix condition number – a measure of the sensitivity of a matrix and of the ability of MPC to effectively decouple the process so that each controlled variable can be maintained at its target. Matrices that have a condition number around 1 are considered well conditioned, while those much greater than 1 are considered ill conditioned.

model predictive control – model-based control of future trajectory of a process variable from the changes of manipulated variables (same as constrained multivariable predictive control)

model-based PCA – a variation of PCA which can be used for monitoring both continuous and batch processes with the help of an accurate process model

move limit – largest change in the manipulated variable per execution of a model predictive controller (normally set large for simulation test but reduced during commissioning)

MSPC – multivariate statistical process control (MSPC) is a statistical tool by which useful information can be obtained from hundreds of process variables with noise filtered out

multiway-PCA – a variation of that focuses on unfolding three dimensional batch data into two dimension before applying PCA algorithm

NIPALS – an iterative algorithm by which one can calculate score and loading matrices

nonlinear system – gain, time constant, or time delay and hence controller tuning settings are not constant but a function of time, direction, operating point, or load

offline monitoring – process variable data are only available at the end of a batch

OLE – Object Linking and Editing

online monitoring – process variables are measured and monitored at regular intervals during the batch

OPC – OLE for Process Control

open loop gain – final % change in the controller input divided by the % change in the controller output with the controller in manual for a self-regulating process. It is dimensionless steady state gain that is the product of the gains for the manipulated variable (e.g., valve gain), the process variable, (i.e., process gain), and controlled variable (e.g., measurement span). It is often simply referred to as the process gain.

open loop time constant – time it takes for the process variable to reach 63 percent of its change after its time delay for a step change in controller output when the controller is in manual. It can originate in the final element, process, or measurement. It is often simply referred to as the process time constant.

oscillation period – time between successive peaks in any type (damped or undamped) of oscillation (the period of damped oscillations is usually larger than the ultimate period)

PC – the uncorrelated vectors from PCA called principal component

PCA – principal component analysis (PCA) is a dimensional reduction method that identifies a subset of uncorrelated vectors so as to capture most of the variation in the data

penalty on error (PE) – an MPC controlled or constraint variable tuning value that determines how much the variable is penalized for errors and violations. The higher the value, the more aggressively the controller provides feedback correction

penalty on move (PM) – a tuning value of an MPC manipulated variable that determines how much the manipulated variable is penalized for moving (also known as move suppression). The higher the value, the slower the moves

PLS – partial least squares (PLS) focuses on maximizing the covariance between the process variable matrix (X) and the quality variable matrix (Y)

positioner – controller mounted on the valve whose set point is desired position and whose controlled variable is shaft position (also known as a digital valve controller)

primary controller – master or outer loop controller whose output is the set point (CAS set point) of a secondary controller in a cascade control system

process action – relative direction of change in process variable for a change in the manipulated variable ("direct" for same direction and "reverse" for opposite direction)

process gain – final change in the process variable divided by the change in the manipulated variable (commonly used as the open loop gain)

process variable – direct or inferential measurement of a variable in the process (normally in process engineering units)

process variable filter – see digital filter

proportional action – controller output changes that are proportional to the change in error (proportional mode of a PID controller)

proportional band – percent change in control error that causes a 100 percent change in controller output (%). It is equivalent to 100 percent divided by controller gain.

rate time – tuning parameter multiplier for the derivative mode of a PID controller (seconds)

remote cascade – mode where the preferred value for the controlled variable comes from model predictive control, sequence, or supervisory controller (RCAS)

remote output – mode where the controller algorithm is suspended and the controller output is set by a discrete action, interlock, or sequence (ROUT)

reset time – tuning parameter divisor for the integral mode of a PID controller (seconds per repeat). It is the inverse of the reset setting given in repeats per second

repeatability – short-term maximum scatter in the output for the same input approached from the same direction and at the same conditions (see reproducibility)

reproducibility – long-term maximum scatter in the output for the same input approached from both directions and at the same conditions (includes repeatability and drift)

resolution – minimum change in the input in the same direction that will cause a change in the output (important for a control valve and a manipulated variable)

runaway process - a process that accelerates for a change in the manipulated variable when the controller is in manual if there are no disturbances

saturation – controller output is at either low or high limits that are normally set to match the valve being shut and wide open, respectively (controller is effectively disabled)

scan time – time interval between successive scans for digital devices (same as update time or control interval and the inverse of scan rate or frequency)

score matrix – T matrix in Eq. 8-2q; a column in the T matrix indicates the relationship between each row of raw data (observations for continuous process)

secondary controller – slave or inner loop controller whose (CAS) set point is manipulated by the output of a primary controller in a cascade control system

self-regulating process – a process that decelerates and reaches a steady state for a change in the manipulated variable when the controller is in manual if there are no disturbances (no integrating or runaway response)

sensitivity – ratio of the steady state change in output to the change in the input (poor sensitivity is caused by poor resolution and a low steady state gain or flat curve)

signal filter – see digital filter

slip – rapid movement of valve trim caused by dynamic changes in friction (the zipping past the desired position means the controller will never reach its set point)

smart transmitter – transmitter with a microprocessor to compensate for changes in ambient and process conditions and to provide remote calibration and diagnostics

SPE **control chart** – a multivariate process control chart in which the raw data are projected to the residual subspace

static gain – see steady state gain

steady state gain – final change in the output of loop components (controller, valve, process, measurement) divided by the change in input (static or zero frequency gain)

stick – lack of valve trim movement for a change in input signal caused by friction or an undersized actuator (major source of a valve's resolution limit)

SVD – in singular value decomposition (SVD), a matrix is decomposed into the product of a diagonal (eigenvalue) and an orthonormal (eigenvector) matrix

time constant – time it takes for the process variable to reach 63 percent of its change after its time delay for a step change.

time delay – time it takes for the process variable to get out of the noise band for a change in the manipulated variable (same as dead time)

time lag – see time constant

time to steady state (T_{ss}) – the time it takes the process variable or controlled variable to reach its final value (or about 99 percent of the change in the process variable) after a change to the manipulated variable.

tuning weight – tuning adjustment in a model predictive controller (when applied to CV or QV it sets priority and when applied to MV it sets move suppression for stability)

UCL / LCL – upper control limit (UCL) and lower control limit (LCL) define the region where a process is in control. If a point lies outside this region, the process is out of control

ultimate gain – controller gain at equal amplitude oscillations for proportional only control (occurs at 180 degree phase shift and is inverse of static gain and amplitude ratio)

ultimate period – period of equal amplitude oscillations for proportional only control (occurs at 180 degree phase shift and is the inverse of the natural period in radians/sec)

update time – time interval between successive updates for digital devices (same as scan time or control interval and the inverse of scan rate or frequency)

upset – see disturbance

valve action – relative direction of change in flow for a change in the input signal to a valve ("direct or inc-open" for fail closed and "reverse" or "inc-close" for fail open)

Appendix B:
Condition Number

Appendix B

Condition Number

Start with a 2x2 steady state gain matrix K:

$$K = \begin{bmatrix} K_{11} & K_{12} \\ K_{21} & K_{22} \end{bmatrix} \tag{B-1}$$

Compute the transpose K^T:

$$K^T = \begin{bmatrix} K_{11} & K_{21} \\ K_{12} & K_{22} \end{bmatrix} \tag{B-2}$$

Compute a new matrix A that is the product of the transpose and the original matrix:

$$A = K^T * K = \begin{bmatrix} K_{11}*K_{11}+K_{21}*K_{21} & K_{11}*K_{12}+K_{21}*K_{22} \\ K_{12}*K_{11}+K_{22}*K_{21} & K_{12}*K_{12}+K_{22}*K_{22} \end{bmatrix} \tag{B-3}$$

Compute the quadratic coefficients for the polynomial characteristic equation $|A - \lambda*I| = 0$ where "|" denotes the determinant, λ is an eigenvalue matrix, and I is the identity matrix:

$$a = 1 \tag{B-4}$$

$$b = (K_{11}*K_{11}+K_{21}*K_{21})+(K_{12}*K_{12}+K_{22}*K_{22}) \tag{B-5}$$

$$c = (K_{11}*K_{11}+K_{21}*K_{21})*(K_{12}*K_{12}+K_{22}*K_{22}) \\ - (K_{11}*K_{12}+K_{21}*K_{22})*(K_{12}*K_{11}+K_{22}*K_{21}) \tag{B-6}$$

Compute the eigenvalues λ_1 and λ_2 as the two roots of the quadratic equation:

$$\lambda_1 = [-b + (b^2 - 4*a*c)] / (2*a) \tag{B-7}$$

$$\lambda_2 = [-b - (b^2 - 4*a*c)] / (2*a) \tag{B-8}$$

Compute the singular values Λ_1 and Λ_2 as the square root of the eigenvalues:

$$\Lambda_1 = \lambda_1^{0.5} \tag{B-9}$$

$$\Lambda_2 = \lambda_2^{0.5} \tag{B-10}$$

Compute the condition number κ as the ratio of maximum to minimum singular values:

$$\kappa = \Lambda_1 / \Lambda_2 \tag{B-11}$$

Appendix C:
Unification of Controller Tuning Relationships

Appendix C

Unification of Controller Tuning Relationships

An intensive search is underway for a unified field theory that would bring together quantum and gravitational forces and provide the underlying truth behind physical laws. In this appendix we show that we are at least fortunate enough to have achieved a unification of Lambda, internal model control, the Ziegler-Nichols reaction curve, and ultimate oscillation tuning methods for bioprocess control. For the control of vessel temperature, concentration, and gas pressure, the controller tuning equations from diverse methods are reducible to a common form, in which the maximum controller gain (equation 3-3f in chapter 3) is proportional to the time-constant-to-dead-time ratio (τ_1 / τ_d) and is inversely proportional to the open loop gain (K_o), commonly known as the process gain. This common form is easy to remember and provides insight into the relative effects of process dynamics on tuning and hence on loop performance. This appendix concludes with a derivation of the equation to predict the control error (integrated absolute error) in terms of the tuning settings (equation 2-2a in chapter 2) from a PI controller's response to load disturbances.

Lambda tuning provides stable results for any Lambda value. Normally, Lambda is set large enough to provide the degree of slowness desired to reduce interaction and promote the coordination of loops. If Lambda is set much smaller than normally expected, as outlined in this section, then the result is the common form, which provides maximum disturbance rejection and minimum integrated absolute error. Thus, Lambda tuning has the advantage of enabling the user to achieve a variety of objectives by setting the degree of transfer of variability from the process output to process input.

The Lambda tuning equations for self-regulating processes are as follows:

$$\lambda = \lambda_f * \tau_1 \tag{C-1a}$$

$$T_i = \tau_1 \tag{C-1b}$$

$$K_c = \frac{T_i}{K_o * (\lambda + \tau_d)} \tag{C-1c}$$

$$T_d = \tau_2 \tag{C-1d}$$

$$K_o = K_{mv} * K_{pv} * K_{pv} \tag{C-1e}$$

The Lambda factor is the ratio of closed loop time constant (figure 3-3a) to the open-loop time (figure 3-3b), where the open-loop time constant is the largest time constant (τ_1). For maximum load rejection capability, a Lambda equal to the total loop dead time ($\lambda = \tau_d$) can be used, and the loop will still be stable if the dynamics are accurately known. For many temperature loops, this corresponds roughly to a 0.1 Lambda factor ($\lambda_f = 0.1$). If we substitute this small Lambda into equation C-1c, we end up with the simplified internal model control (SIMC) equation (see equation C-1f [equation 3-3f in chapter 3]). This equation has recently been documented as providing the best load rejection (tightest control) for self-regulating processes [1].

Until recently, proportional band (PB) was predominantly used rather than controller gain. The equation for minimum proportional band is 100 percent divided by equation C-1f since the proportional band is 100 percent divided by the controller gain (PB =100% / K_c). Proportional band is the percent change in the control error that will cause a 100 percent change in controller output from the proportional mode.

$$K_c = 0.5 * \frac{\tau_1}{K_o * \tau_d} \qquad (C\text{-}1f)$$

Equation C-1f, which is equation 3-3f in chapter 3, has dominated the literature since the days of Ziegler and Nichols. The multiplier ranges from 0.4 to 0.8, and the exponent of the time-constant-to-dead-time ratio varies from 0.9 to 1.0. The differences in the multipliers or exponents are insignificant because in practice the user has backed off from the maximum gain, and the effect of errors or changes commonly seen in the identified process gain, time constant, and dead time is larger than the effect of the coefficients.

If we substitute the definition of a pseudo or "near" integrator gain per equation C-1g (equation 3-3g in section 3-3) into equation C-1f we end up with the equation C-1h, which is the SIMC tuning that has been shown to provide the tightest control of integrating processes [1].

$$K_i = \frac{K_o}{\tau_1} \qquad (C\text{-}1g)$$

$$K_c = 0.5 * \frac{1}{K_i * \tau_d} \qquad (C\text{-}1h)$$

If we change nomenclature such that the integrating process gain is the reaction rate ($K_i = R$) and the dead time is the delay time ($\tau_d = L$), then we end up with the controller gain per the Ziegler-Nichols "reaction curve" method developed in the 1940s.

The "reaction curve" method is suitable for self-regulating processes that have large time constants and integrating processes. However, the documentation of the "reaction curve" method showed that the process was lined out ($CV_1 / \Delta t = 0$) just before the change is made in the controller output. As a result, R could be computed from just the ramp rate of the process variable after the change. This may not be the case, especially for integrating processes, since the controller is in manual (see figure 2-3a in section 2-3). Therefore, it is critical to take into account the change in ramp rates from "before" to "after" the change and to use percent rather than engineering units, as shown in equation C-1i. The "short cut" method presented in the book *Good Tuning: A Pocket Guide, 2nd Edition* provides a detailed procedure for using equation C-1i and correcting the observed dead time for the effect of final element resolution or dead band so as to quickly estimate the controller tuning for slow processes [2].

$$K_i = \frac{CV_2/\Delta t - CV_1/\Delta t}{\Delta CO} \tag{C-1i}$$

It may be difficult to accurately identify the second-largest time constant. Therefore, the internal model control computation of the derivative time as shown in equation C-1j may be useful. For a first-order (single time constant) approximation of a concentration or temperature response, about half of the total loop dead time originates from the second-largest time constant, as shown in equation C-1k. Since in this case the time constant is much larger than the dead time, the dead-time term in the denominator of equation C-1j becomes just twice the largest time constant ($2*\tau_1$). If you then cancel out the time constant in the numerator and denominator, you end up with the derivative time being approximately equal to half the dead time. Equation C-1j coupled with equation C-1k is reducible to equation C-1d.

$$T_d = \frac{\tau_1 * \tau_d}{2 * \tau_1 + \tau_d} \tag{C-1j}$$

For interactive lags found in temperature, concentration, and gas pressure processes:

$$\tau_2 = 0.5 * \tau_d \tag{C-1k}$$

The Lambda tuning equations for integrating processes are as follows:

$$\lambda = \lambda_f / K_i \tag{C-1l}$$

$$T_i = 2 * \lambda + \tau_d \tag{C-1m}$$

$$K_c = \frac{T_i}{K_i * (\lambda + \tau_d)^2} \tag{C-1n}$$

The Lambda factor is the ratio of closed loop arrest time to open-loop arrest time where the open loop arrest time is simply the inverse of the integrating process gain ($1/ K_i$). For maximum load rejection capability, a Lambda equal to the total loop dead time ($\lambda = \tau_d$) can again be used, and the loop will still be stable if the dynamics are accurately known. In this case, equation C-1n reduces to equation C-1h, but with a multiplier of 0.75 instead of 0.5. The higher multiplier is insignificant since it is rarely desirable, and dynamics are seldom known accurately enough to take advantage of this increase in the controller gain.

For many processes that have a true or near-to-true integrating response, the controller gains computed by equation C-1h are much higher than desired or needed. This is particularly the case for bioreactors since the disturbances are so slow. However, a much lower controller gain can lead to nearly sustained oscillations with a very long period. To prevent this from occurring, equation C-1o, which is developed from the transfer function for the closed loop response of an integrating process, can be used to ensure that the response is over damped.

The integral time needed to ensure an over-damped response in integrating processes is as follows:

$$T_i > \frac{4}{K_i * K_c} \tag{C-1o}$$

For self-regulating processes, the transfer function for the closed loop response yields equation 3-3p for an over-damped response. The product of the controller gain and the process gain is much larger than one for temperature and concentration loops on well-mixed volumes because the time constant is so large, which means you can cancel out the product of gains in the numerator. If you then use equation C-1g to get an equivalent process integrating gain, then equation C-1p is reducible to equation C-1o.

The integral time needed to ensure an over-damped response in self-regulating processes is as follows:

$$T_i > \frac{4 * (K_o * K_c * \tau_1)}{(1 + K_o * K_c)^2}$$ (C-1p)

Ziegler and Nichols developed controller tuning equations based on field measurements of the ultimate gain and ultimate period. For a manual tuning test, the derivative time is set to zero, and the integral time is set at least ten times larger than normal so most of the controller response is from the proportional mode. The controller gain is then increased to create equal sustained oscillations. The controller gain at this point is the ultimate gain, and the oscillation period is the ultimate period. In industry, the gain is only increased until decaying oscillations first appear to reduce the disruption to the process. Auto tuners and adaptive controllers make this manual controller tuning unnecessary. The "relay method" is extensively employed by "on-demand" auto tuners to automatically compute the ultimate period and gain by switching the controller output when it crosses and departs from a noise band centered on the set point [2] [3].

The ultimate gain for self-regulating processes using the amplitude ratio is [3]:

$$K_u = \frac{[1 + (\tau_1 * 2 * \pi/T_u)^2]^{0.5}}{K_o}$$ (C-1q)

For the Ziegler-Nichols ultimate oscillation method, the controller gain is simply a fraction of the ultimate gain, and the integral (reset) time is a fraction of the ultimate period, as the following equation shows for a PI controller:

$$K_c = 0.4 * K_u$$ (C-1r)

$$T_i = 0.8 * T_u$$ (C-1s)

If you also take into account that the squared expression in the numerator of equation C-1q is much larger than one, you end up with equation C-1t. For most temperature and composition loops on bioreactors, the ultimate period is approximately four times the dead time ($T_u = 4*\tau_d$). If you substitute this relationship into C-1t, you end up with equation C-1u.

After multiplying numerical factors, equation C-1u becomes equation C-1v (equation C-1f with a slightly larger multiplier).

$$K_c = 0.4 * \frac{(\tau_1 * 2 * \pi)/T_u}{K_o} \tag{C-1t}$$

$$K_c = 0.4 * \frac{(\tau_1 * 2 * \pi)/(4 * \tau_d)}{K_o} \tag{C-1u}$$

$$K_c = 0.6 * \frac{\tau_1}{K_o * \tau_d} \tag{C-1v}$$

If the ultimate period is about four times the dead time ($T_u = 4*\tau_d$), then the integral time ends up as about three times the dead time, per equation 3-1s for the ultimate oscillation method. This is the same result you get according to equation C-1m for the Lambda tuning method when Lambda is reduced to equal the dead time. This reset time is generally considered to be too fast. The SIMCA method states that while four times the dead time provides the best performance and that an increase to eight times the dead time provides better robustness. If four times the dead time is used, then equations C-1d and C-1k result in a reset time that is eight times the rate time setting. Though most of the literature shows the rate time as being equal to one-quarter the reset time, in practice a rate time that is one-eighth to one-tenth the reset time provides a smoother response.

Equation 2-2a in chapter 2 for the integrated absolute error (IAE) can be derived from a PI controller's response to a load upset. The module execution time (Δt) is added to the reset or integral time (T_i) to show the effect of how the integral mode is implemented in some digital controllers. An integral time of zero ends up as a minimum integral time equal to the execution time so there is not a zero in the denominator of equation C-1w. For analog controllers, the execution time is effectively zero [6].

$$\Delta CO = K_c * \Delta E_t + [K_c / (T_i + \Delta t)] * \text{Integral } (E_t * \Delta t) \tag{C-1w}$$

The errors before the disturbance and after the controller has completely compensated for the disturbance are zero ($\Delta E_t = 0$). Therefore, the long-term effect of the proportional mode, which is first term in equation C-1w, is zero. Equation C-1w reduces to equation C-1x [5].

$$\Delta CO = [(K_c / (T_i + \Delta t)] * \text{Integral } (E_t * \Delta t) \tag{C-1x}$$

For an over-damped response:

$$IAE = Integral\ (E_t * \Delta t) \tag{C-1y}$$

The open loop error is the peak error for a step disturbance in the case where the controller is in manual (loop is open). The open loop error (E_o) is the open loop gain (K_o) times the shift in controller output (ΔCO) required to compensate for the disturbance when the controller is in auto (loop is closed).

$$E_o = K_o * \Delta CO \tag{C-1z}$$

Equation C-1x solved for the IAE defined in equation C-1y and the open loop error defined in equation C-1z becomes equation C-1aa. If you ignore the effect of module execution time (Δt) on the integral mode, equation 2-2a in chapter 2 is equation C-1aa for an over-damped response because the integrated absolute error (IAE) is the same as the integrated error (E_i). Even for a slightly oscillatory response, the approximation has proven to be close enough [4].

$$IAE = \frac{1}{(K_o * K_c)} * (T_i + \Delta t) * E_o \tag{C-1aa}$$

For the control of vessel temperature, concentration, and pressure, we can use equation C-1f for the maximum controller gain and four times the dead time for the minimum reset time to express the minimum integrated absolute error in terms of the process dynamics. The resulting equation C-1ab shows that the minimum integrated absolute error is proportional to the dead time squared for tuning settings that give the tightest control. Note that the open loop gain has cancelled out. This equation can be independently derived by multiplying the peak error for a step disturbance by the dead time [3] [6].

$$IAE = 2 * \frac{\tau_d}{\tau_1} * (4 * \tau_d + \Delta t) * E_o \tag{C-1ab}$$

In practice, controllers are not tuned this aggressively. Often the reset time is set equal to the time constant and a Lambda factor of 1.0 is used, which corresponds to a controller gain that is about ten times smaller than the maximum controller gain for a bioreactor's primary loops (e.g., composition, pressure, and temperature).

If the disturbance is not a step change, the integrated absolute error will be smaller. The effect of a slow disturbance can be approximated by adding the disturbance time constant to the open loop time constant (τ_1) in the denominator of equation C-1ab.

An increase in the module execution time shows up as an increase in the loop dead time for unmeasured disturbances. If the disturbance arrives immediately after the process variable is read as an input to the module, the additional dead time is about equal to the module execution time. If the disturbance arrives immediately before the process variable is read, the additional dead time is nearly zero. On the average, the additional dead time can be approximated as 50 percent of the module execution time. Simulations that create a disturbance that is coincident with the controller execution will not show much of an effect of execution time on performance. This scenario misleads users into thinking that the execution time of model predictive control is not important for load rejection. For chromatographs where the result is only available for transmission after the processing and analysis cycle, the additional dead time is 150 percent of the analyzer cycle time [2] [3] [5].

Equation C-1ab shows the effect of the largest time constant, loop dead time, and module execution time on absolute integrated error if the controller is always retuned for maximum performance. A detuned controller may not do much better than a tightly tuned controller for a larger loop dead time or module execution time [5]. Thus, the value of reducing these delay times depends on the controller gain used in practice. For example, the controller gain is simply the inverse of the open loop gain for a Lambda factor of one in a loop with a dead time much smaller than the time constant. In other words, equation C-1c reduces to equation C-1ac.

$$K_c = \frac{1}{K_o} \qquad\qquad \text{(C-1ac)}$$

If you substitute equation C-1ac into equation C-1f and solve for the dead time, you end up with equation C-1ad. This shows a Lambda factor of one on a primary reactor loop, which implies a dead time that is about ½ of the time constant. The integrated absolute error for this case will not appreciably increase until the dead time is about ½ of the time constant. Thus, time and money spend on reducing the dead time or module execution time below this implied dead time has little value unless the controller is retuned [5].

$$\tau_d = 0.5 * \tau_1 \qquad\qquad \text{(C-1ad)}$$

Nomenclature

ΔCO	=	shift in controller output to compensate for disturbance (%)
E_i	=	integrated error (% seconds)
E_o	=	open loop error for a step disturbance (%)
E_t	=	error between SP and PV during the disturbance
IAE	=	integrated absolute error from the disturbance (% seconds)
K_c	=	controller gain (dimensionless)
K_i	=	integrating gain (%/sec/% or 1/sec)
K_o	=	open loop gain (dimensionless)
K_u	=	ultimate gain (dimensionless)
λ	=	Lambda (closed loop time constant or arrest time) (sec)
λ_f	=	Lambda factor (ratio of closed to open loop time constant or arrest time) (dimensionless)
Δt	=	module execution time (sec)
τ_d	=	total loop dead time (sec)
τ_1	=	largest open loop time constant (sec)
τ_2	=	second largest open loop time constant (sec)
T_i	=	integral (reset) time setting (sec/repeat)
T_d	=	derivative (rate) time setting (sec)
T_u	=	ultimate period (sec)

References

C-1.   Skogestad, Sigurd, "Simple Rules for Model Reduction and PID Controller Tuning", *Journal of Process Control*, 13, (2003): 291-309.

C-2.   McMillan, Gregory K., *Good Tuning: A Pocket Guide, 2d ed.*, ISA, 2005.

C-3.   Blevins, Terrence L., McMillan, Gregory K., Wojsznis, Willy K., and Brown, Michael W., *Advanced Control Unleashed: Plant Performance Management for Optimum Benefits*, ISA, 2003.

C-4.   Shinskey, F. G., *Feedback Controllers for the Process Industries*, McGraw-Hill, 1994.

C-5.   McMillan, Gregory, K., *Tuning and Control Loop Performance, 3d ed.*, ISA, 1991.

C-6.   Shinskey, F. G., "The Effect of Scan Period on Digital Control Loops", *InTech*, June 1993.

Appendix D:
Modern Myths

Appendix D

Modern Myths

Myth 1 – You need an advanced degree to do advanced control. Not so anymore. New software packages that can be used to design a virtual plant automate much of the required expertise and eliminate the need for special interfaces. The user can now focus mostly on the application and the goal.

Myth 2 – Dynamic simulations and model-based control are only applicable to continuous processes. Since most of the applications are in the continuous industry, this is a common misconception. While it is true that a steady state simulation is not valid since there is by definition no steady state in batch, dynamic simulation can follow a batch as long as the software can handle zero flows and empty vessels. Model-predictive control (MPC), which looks at trajectories, is suitable for the real-time optimization (RTO) of fed batch processes. The opportunities MPC provides to improve a process's efficiency add up to be about 25 percent for batch compared to 5 percent for continuous operations.

Myth 3 – You need consultants to maintain dynamic simulations and model-based control systems. No longer true. The ease of use of new software allows the user to get much more involved, which is critical to ensuring that the plant gets the most value out of the models. Previously, the benefits started to diminish as soon as the consultant left the job site. Now, the user should be able to tune, troubleshoot, and update the models.

Myth 4 – You don't need good operator displays and training if you have well-designed RTO and MPC systems. The operators are the biggest constraint in most plants. Even if the models used for real-time optimization and model-based control are perfect, operators will take these systems off line if they don't understand them. The new guy in town is always suspect, so the first time an operational problem occurs and no one is around to answer questions, the RTO and MPC systems are turned off—even if they are doing the right thing. Training sessions and displays should focus on showing the individual contribution of the trajectories for each controlled, disturbance, and constraint variable in relation to the observed changes in the manipulated and optimization variables.

Myth 5 – Simple step (bump) tests are never enough. You must do a PRBS test. A complete PRBS test may take too long. The plant may have moved to an entirely different state, tripped, or—in the case of a batch operation—finished, before a PRBS test is complete. As a minimum, one step in each direction should be held to steady state. The old rule is true: if you can see the model from a trend, it is there. Sometimes, the brain can estimate the process gain, time delay, and lag better than a software package.

Myth 6 – You need to know your process before you start a RTO or MPC application. This would be nice, but often the benefits provided by a model stem from the knowledge discovered during the systematic building and identification procedures. Frequently, the understanding gained by developing models leads to immediate benefits in terms of better set points and instruments. The commissioning of the RTO and MPC is the icing on the cake and locks in benefits for varying plant conditions.

Myth 7 – Optimization by pushing constraints will decrease on-stream time. Not true. MPC and RTO recognize future violations of constraints for unforeseen problems and will back off from the edge to increase on-stream time.

Appendix E: Enzyme Inactivity Decreased by Controlling the pH with a family of Bezier Curves [1]

Appendix E

Enzyme Inactivity Decreased by Controlling the pH with a family of Bezier Curves [1]

Michael J. Needham, Validation Engineer
ROVISYS
Aurora, Ohio

A client asked if it was possible to better control the inactivity of enzymes that were pH sensitive. The question arose because the client needed a stable, or possibly increasing, microbial growth environmental that would prevent or shift the biosynthesis capabilities of the process harvest. The biosynthesis required an essential active enzyme concentration in a cell culture growth medium. Reduced biosynthesis capabilities, caused by an inactive enzyme, posed an economic risk for the client.

The biosynthesis of pharmaceutical drug products produced through biotechnical manufacturing processes begins in a laboratory. The informational components essential to this development include charts showing the production capability of the drug of interest compared with pH, time, substrate concentration, enzyme concentration, and the like. These informational components are the basis and starting point for the innovative enzyme controller (IEC) to be developed. To understand how these laboratory charts fundamentally improve the approach to process control, this appendix begins with a brief review of the methods used by physical biochemists to analyze these types of reactions.

The biological cellular growth mediums that make up enzyme-catalyzed reactions are difficult to work with analytically for the following key reasons [2]:

- To achieve lower activation energies for a reaction, enzymes require extremely low concentrations compared to their substrate. These low concentrations as well as the complexity of enzyme-substrate binding make it difficult to measure what the enzyme is actually doing.

- The kinetic constants of complex enzymes cannot be determined by steady-state experiments. In a step reaction sequence to produce a particular drug there are often at least six kinetic constants and more are possible (see figure E-1).

- The molecular dissociation in these biological mediums is difficult to determine.

- Conventional steady-state equilibrium methods are not linearly applicable to enzymes.

- The activities of many enzymes vary with pH in the same way that simple acids and bases ionize. Enzymes involved in the sequential reactions associated with an microorganism's biosynthetic pathway growth kinetics are very complex, diverse, and huge (molecular weights can be as high as 1,000,000 Daltons).

- The activity of enzyme reaction velocities can be inhibited by reactive competition from other components in the medium or even the substrate [3].

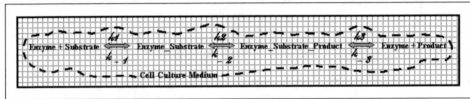

Figure E-1. Reaction Mechanism

Recently, physical biochemists have derived a standard state where the activities and concentrations normally used to analyze reaction kinetics are not required. Hammes developed an analytical method called "Relaxation Methods" to study fast chemical reaction components at equilibrium. The Hammes method can be briefly described as follows:

A chemical system at equilibrium is suddenly perturbed in such a way that it finds itself out of equilibrium under the new conditions. It then relaxes to a state of equilibrium under the new conditions, and the rate of relaxation is governed by the rate constants and mechanism of the process.

In this appendix we will use relaxation method analysis and steady-state theory to describe the physical model of an enzyme-catalyzed reaction as shown in figure E-1. Our precepts are as follows:

- At equilibrium, the rates of forward and back reactions are the same:

    k1 = k-1
    k2 = k-2
    k3 = k-3

- Individual rate constants cannot be determined, only their ratio.

- Many reaction properties undergo constant and rapid fluctuations around some average or median state. Any long-term measurements yield the same result if the system is truly at equilibrium.

- Relaxation kinetics give simple first-order exponential time regardless of how many molecules are involved as either reactants or products.

- After the measurement is analyzed, the concentrations charted over time should describe a straight line.

- Steady state is an approximation since the substrate is gradually being depleted. However, this approximation will be very good as long as the rate measurements are restricted to a short time interval during which the substrate's concentration does not greatly change.

Departing from conventional control strategies, the principle object of the IEC will be:

Maintaining the equilibrium by controlling the reaction perturbation as defined by the laboratory charts, which is technically supported by the Relaxation Method's physical model for the enzyme-catalyzed reaction of interest.

To divide the laboratory charts into different reactant intermediate states, we will use the transition state theory [4] (see figure E-2). This theory is supported by the Hammond postulate for Gibbs free energy of mathematical maximums and minimums for reaction growth diagrams. Stated briefly, the transition state theory for enzyme-catalyzed chemical reactions is as follows:

- Transition states occur at mathematical maximums in reaction growth diagrams for a reaction, and intermediates occur at mathematical minimums.

- At mathematical maximums on reaction growth diagrams, chemical bonds are in the process of being made and broken.

- At mathematical minimums on reaction growth diagrams, intermediates have fully formed chemical bonds.

- The reaction growth diagram plots the energy of the reagents (we use both growth and pH) as the reaction proceeds.

Briefly stated, the portion of the Hammond postulate of interest is as follows:

- If there is an unstable intermediate in the chemical reaction pathway, the reaction's transition state will resemble the structure of the intermediate. The reasoning is that the unstable intermediate will be a mathematical maximum in the reaction coordinate diagram.

- Gibbs free energy curves for a substrate and product intersect on the reaction coordinate at the position of the transition state. A change of structure away from the seat of reaction destabilizes it, displacing its energy curve higher. The point of intersection moves closer to the substrate.

The importance of transition state theory is that it relates the rate of a reaction to the difference in Gibbs energy between the transition state and the ground state.

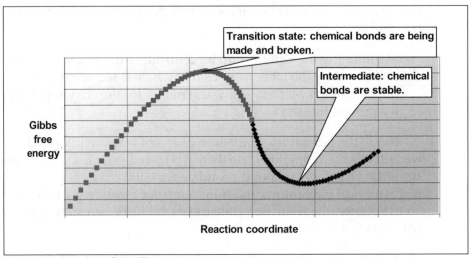

Figure E-2. Transition State Theory

We approached the selection of the nonlinear control strategy conceptually [5]. As Antje Kann has noted [6], "models are often developed…with knowledge of theory and algorithms, but who may not be intimately familiar with the users' situations. Modeling choices made … are often guided by practical concerns such as solvability, run-time issues, and ease of implementation in code. However, these choices may impair the ease with which a model can be applied to solve the users' problems." Following this train of thought, we took a fresh look. The laboratory charts would be the best starting point for the bases of the controller development. Well-documented analysis would be available for all the process dynamics, measured variables, responses, and also unanswerable process questions for the drug of interest. Classical multivariable, nonlinear control strategies also represented well-documented information. Also, a good curve-generating engine would help turn data from the laboratory charts into information.

Parametric equations allow curves to have the unusual ability to move straight up or even curve back and cross themselves (see figure E-3). Cubic Bezier curves, which belong to the family of parametric curve equations, were selected as the first set of parametric curves to use because they do not need a tangent vector. Bezier curves are characterized by defining four coordinate points (origin, destination, values of X, and values of Y), one linear variable coordinate point (t), and two polynomial equations. The two polynomial equations are used to calculate the values of X and Y between an increasing third variable (t).

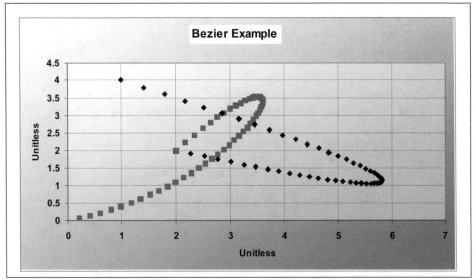

Figure E-3. Example of a Parametric Curve Crossing Itself Twice

The following is a simple example illustrating the development of a cubic Bezier curve [7]:

Two polynomial equations define the coordinate points on the curve. Both are evaluated for an arbitrary number of values of (t). In our example we will increment as follows:

- $t = t + 0.02, t \rightarrow 1$ (that is, 50 values of t)

As increasing values for (t) are supplied to the equations, the point defined by $x\,(t), y\,(t)$ moves from the origin to the destination:

- $x(t) = ax t^3 + bx t^2 + cx t + x0$

  $x1 = x0 + cx/3$
  $x2 = x1 + (bx + cx)/3$
  $x3 = x0 + ax + bx + cx$

- $y(t) = ay t^3 + by t^2 + cy t + y0$

  $y1 = y0 + cy/3$
  $y2 = y1 + (by + cy)/3$
  $y3 = y0 + ay + by + cy$

The coefficient values a, b, and c can be determined using the following three equations:

- $cx = 3(x1 - x0)$

- $bx = 3(x2 - x1) - cx$

- $ax = x3 - x0 - cx - bx$

First, let us select four control cubic coordinates at random:

- origin: xp0 = 3.5 and yp0 = 0

- control point 1: xp1 = 3.5 and yp1 = 1

- control point 2: xp2 = 5 and yp2 = 6

- destination: xp3 = 6 and yp3 = 3

These four coordinates can be evaluated as coefficients:

- cx = 0

- cy = 3

- bx = 4.5

- by = 12

- ax = -2

- ay = -12

Then, calculating the first coordinate pair at t = 0.02:

- x(0.02) = 3.50

- y(0.02) = 0.06

For $t = 0.04$:

- $x(0.02) = 3.50$

- $y(0.02) = 0.14$

For $t = 0.06$, and so on.

Figure E-4 graphically demonstrates the dramatically different results obtained from two selections for control coordinates.

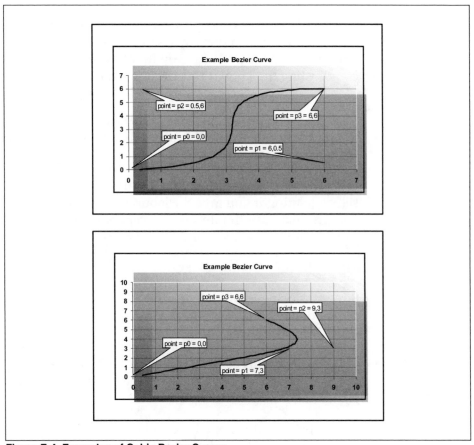

Figure E-4. Examples of Cubic Bezier Curves

We used loosely coupled data from the laboratory charts and borrowed concepts from classical controllers with multivariable [8], nonlinear capabilities that are generally used for pH regulation. This approach produced the following filtered list of controller types:

- Endpoint Control (EpC)

  - Strengths: Stoichiometrics

- How IEC uses this concept: The stoichiometrics associated with enzyme-substrate reactions that occur in cell culture growth mediums are much more complex, both physically and chemically. We will use measured variables as an indication value and as associated other parameters before we determine the value of the final control element.

- Gain Compensation Control (GCC)

  - Strengths: Can function with nonlinear control measurements in real time.

  - How IEC uses this concept: We reduced or in some cases eliminated controller response by giving the laboratory charts for pH, time, and the product growth rate's Bezier capability and by adjusting the known process nonlinearities to be predictable.

- Adaptive Control (AC)

  - Strengths: Controller mode values change with the dynamics of the measured variables.

  - How IEC uses this concept: The controller and process will be self-regulating if you give the laboratory charts for pH, time, and product growth rate's Bezier capability and adjust the controller output to known measurement dynamics.

- Model Predictable Control (MPC)

  - Strengths: Able to predict near future variable values based on the model developed by the pseudo random process variation (PRPV) studies.

  - How IEC uses this concept: A very definitive model is available if you use the laboratory charts as the bases for the controller algorithm,.

The IEC can now be fully developed using the following properties, theories, methods, control strategies, and states for the drug of interest:

- Laboratory informational charts developed for the drug of interest.

- Physical biochemical model precepts by Hammes' Relaxation Method

- Transition state theory with the Hammond postulate

- Steady-state dynamics

- Pseudo steady state

- Endpoint control (EpC) concept of stoichiometrics

- Gain compensation control (GCC) concept for nonlinear measurements

- Adaptive control (AC) concept of mode changes with measured dynamics

- Model predictable control (MPC) concept of near-future predictions for process variables

The following detail requirement serves as the basis for the controller:

- Three laboratory-determined sets of data for product growth rate versus pH, growth rate versus time, and growth rate per pH versus time will be experimentally developed.

- The three sets of laboratory data for product growth rate will be formulated using cubic Bezier control.

- EpC has control components that control the proportional feeding (stoichiometric) of reactive agents in either correct or excess proportions. The IEC will calculate the standard deviation of the laboratory's experimental data and determine if the batch has either a correct or excess proportion of reactive agents.

- GCC has a control component that varies the loop-gain compensation with nonlinear compositions. The IEC will calculate the required loop gain based on the time interval from the state of the batch, pH, and estimated growth rate.

- AC has control components that change controllers' control modes based on the IEC's active control mode. The IEC will change the slave PID controller modes and values based on the following:

- Set-point limits based on the standard deviation of the active control mode

- The integral set by the active median value.

- Derivative set the active estimated growth rate.

- AUTOMATIC/MANUAL/REMOTECASCADE based by operator selection.

- MPC has a control horizon component that uses the model developed for the various coupled control variables to estimate a future value. The growth rate curve determined by experimental methods in the laboratory will be the model.

- Laboratory-determined data point coordinates would be formulated using cubic Bezier control.

- The following measurements will be made and their values compared with the desired values:

  - Time batch started

  - Real-time clock reset to zero at time batch started

  - pH

  - Gain (pH desired at time x/pH actual at time x or growth rate at pH actual at time x/growth rate estimated at time x)

The functional specifications for the IEC that are to be developed are as follows:

- pH measured and displayed on a screen with historical capability

- pH measured data given cubic Bezier capability and displayed on a screen with historical capability

- Chemical transition states, each displayed on a screen

- Chemical transition states data given cubic Bezier capability

- Chemical intermediates, each displayed on a screen

- Chemical intermediates data given cubic Bezier capability

- Real-time growth curve for the microorganism of interest and displayed on two screens, each with historical capability

- Growth rate curve data will be linked to either a chemical transition state or chemical intermediate

- Estimated growth rate curve screens will have a real-time clock displayed linearly

- pH measured data will be linked to each growth rate curve screen

- pH measured data gain will be linked to each growth rate curve screen

- Cubic Bezier control capability for all data sets with historical capability

References

E-1.   This appendix is an excerpt from the article, "Enzyme Decreased by Controlling the pH with a Family of Bezier Curves," published in the May 2006 issue of *Pharmaceutical* magazine.

E-2.   Tinoco Jr., Ignacio, Sauer, Kenneth, and Wang, James C., *Physical Chemistry Principles and Applications in Biological Sciences,* 2d ed., Englewood Cliffs, NJ: Prentice-Hall, 1985.

E-3.   Champe, Pamela C., Harvey, Richard A., and Ferrier, Denise R., *Biochemistry.* Lippincott Williams & Wilkins, 2005.

E-4.   Fersht, Alan, *Structure and Mechanism in Protein Science.* New York: W. H. Freeman, 2000.

E-5.   *ISA Standards Library for Automation and Control, 1st ed.* Research Triangle Park, NC: ISA, 2006.

E-6.   Conversations with Antje Kann, President, Analytics Interactive.

E-7.   Bezier Curves. http://www.math.ubc.ca/~cass/gfx/bezier.html.

E-8.   McMillan, Gregory K., *Tuning and Control Loop Performance.* 2d ed. ISA, 1990.

Index